HEALTHCARE STEWARDSHIP

A 21st Century Global Call to Action

DALE J. BLOCK

Healthcare Stewardship:
A 21st Century Global Call to Action

ISBN: 979-8-89633-055-4 (sc)
ISBN: 979-8-89633-056-1 (e)

PAGE
SOLUTIONS

Page Solutions
541 Buttermilk Pike
Crescent Springs, KY 41017

Printed in the United States of America

CONTENTS

CONTENTS

DEDICATION

I dedicate this book to my wife, Ellen; my three sons, Aaron, Jeremy, and Stuart; their respective spouses, John, Valerie, and Christina; and my grandchildren, Jude, Emilia, and Jax. I also dedicate this book to the memory of my daughter, Julie, whose influence has been a constant inspiration throughout my medical career, motivating me to continually strive to educate the next generation of healers.

PREFACE.
Primum non nocere

Primum non nocere, or *first, do no harm*, is a core ethical principle in healthcare that underscores the commitment to preventing harm in patient care. This principle helps guide healthcare professionals to carefully evaluate the potential risks and benefits of any intervention or treatment. By adhering to *primum non nocere*, clinicians ensure that their actions prioritize patient quality of care and minimize unnecessary safety risks. It promotes a careful, empirically-driven evidence-based approach to critical decision-making, problem solving, and logical reasoning, emphasizing the importance of avoiding clinical interventions that could lead to adverse effects or complications. This principle fosters transparency, trust, truthfulness, and confidence between patients and providers, reinforcing the ethical foundation of medical practice.

Healthcare stewardship is defined as the responsible management of limited healthcare resources to ensure the most effective and ethical delivery of healthcare goods and services. It encompasses principles of efficiency, sustainability, and fairness, ensuring that the limited healthcare resources are used wisely and equitably to maximize patient benefit. *Primum non nocere* aligns closely with healthcare stewardship, as both principles focus on minimizing harm and optimizing quality and outcomes with the appropriate use of limited resources. Stewardship, alone, involves not only managing physical resources but also considering the broader impact of healthcare decisions on patient safety, quality of care, and overall system efficiency. By integrating *primum non nocere* with effective stewardship principles and practices, healthcare providers can enhance the quality and safety of delivered care, lower the total cost of that delivered care, and overall, ensure the sustainability of evidence-based healthcare services delivered around-the-world, ultimately leading to better patient outcomes and a more equitable healthcare system.

Healthcare Stewardship: A 21st Century Global Call to Action explores transformational shifts in the global healthcare delivery system, envisioning a sustainable, integrated, and holistic framework grounded in empirically-driven evidence-based clinical practices. By prioritizing foundational principles of stewardship and fostering capacity-building improvements, this book advocates for sweeping reforms guided by the timeless ethical and moral principle of *primum non nocere* —to first do no harm. It calls for an essential paradigm shift to address urgent challenges, driving innovation and collaboration whilst establishing a resilient global healthcare model for the 21st century.

Healthcare Stewardship stands as the first definitive text to incorporate the principles and practices of stewardship—a concept with deep roots in religious traditions dating back to biblical times—into the management and delivery of scarce healthcare resources. At its core, stewardship involves utilizing limited healthcare goods and services in ways that are clinically, ethically, politically, environmentally, and socially responsible. While this principle may appear straightforward, achieving simplicity within the global healthcare delivery system remains an elusive goal. Layered bureaucracies across regional, national, state, and local levels have rendered healthcare services increasingly complex, inefficient, and ineffective. This book seeks to ignite a sense of urgency—a burning platform—for a transformational paradigm shift, aiming to reshape the global healthcare system into an environmentally sustainable, equitable model of care delivery easily accessible to all.

The purpose of *Healthcare Stewardship* is to raise awareness among all those impacted by the global delivery of healthcare services. This book aims to shed light on the critical challenges, pressing issues, and untapped opportunities within the framework of the global healthcare system. By fostering a shared understanding and commitment to progress, it seeks to engage healthcare professionals, policymakers, patients, and the broader public. The goal is to inspire collective responsibility and catalyze meaningful change, paving the way for a more efficient, equitable, and environmentally sustainable healthcare delivery model worldwide.

Dale J. Block, MD, MBA

Spring 2025

1.0

21st Century Existential Threats.

1.1 Existential Threats to the Global Healthcare Delivery System in the 21st Century.

The global healthcare delivery system faces several existential threats in the 21st century, with one of the most pressing being the rising cost of healthcare (Lopes et al., 2024). High-cost expenses, in both high-income (HIC) and low-middle income LMIC) countries are a significant barrier for many individuals, leading to a substantial number of marginalized and vulnerable people being uninsured, under-insured, and financially-ruined. This financial strain affects not only patients but also healthcare providers and institutions, which struggle to balance high operating costs with the need to deliver high-quality care. The unsustainable cost trajectory threatens to widen health disparities, reduce access to essential healthcare services, and undermine the overall stability of the global healthcare delivery system.

Another existential threat is the increasing burden of chronic diseases, which account for the majority of healthcare spending around-the-world (Chan et al., 2020; Lai et al., 2021; Vodovotz et al., 2020). Conditions such as diabetes, heart disease, cancer, dementia, and obesity are becoming more prevalent due to lifestyle and behavioral factors and an aging global population. Managing these diseases requires long-term, coordinated, comprehensive and complex clinical care requiring significant resources, putting a strain on an already overstretched global healthcare system. Without effective strategies to prevent and manage chronic illnesses, the system risks being overwhelmed, compromising care quality, safety, and accessibility.

The global healthcare workforce crisis is another critical issue facing the global healthcare delivery system (de Vries et al, 2023). There is a growing shortage of healthcare professionals, including doctors, nurses, and allied health workers, exacerbated by an aging workforce and high rates of burnout and job dissatisfaction, due to the recent COVID-19 pandemic. This shortage impacts the delivery of care, leading to longer wait times, reduced patient access, and increased pressure on existing staff, which can compromise patient safety and outcomes. Addressing workforce development, recruitment, and retention is essential to maintaining a robust, responsive, and resilient global healthcare system capable of meeting current and future demands.

Additionally, the integration of advanced medical and information technology with data privacy and cybersecurity concerns presents a significant challenge (Li et al., 2021; Okafor et al., 2023). While advancements in digital health technologies, such as electronic health records (EHRs) and telemedicine, have the potential to improve care delivery, they also introduce vulnerabilities related to data privacy and cybersecurity. The increasing frequency, magnitude, and sophistication of cyberattacks on healthcare systems world-wide pose a threat to patient information and overall healthcare system integrity. Ensuring robust cybersecurity measures and protecting patient data are critical to maintaining trust in the global healthcare system and safeguarding against disruptions that could impede care delivery.

Political and regulatory instability poses a continuous threat to the global healthcare delivery system (Thomas et al, 2020; Berman et al., 2022). Frequent changes in healthcare public policy, funding, and regulations with every country's political cycle create an environment of uncertainty for providers, insurers, and patients. The politicization of healthcare issues can lead to abrupt policy shifts that disrupt the stability and predictability needed for effective long-term planning and investment in healthcare infrastructure. Ensuring a stable and supportive regulatory environment is crucial for fostering innovation, improving care delivery, and addressing the complex challenges facing the global healthcare system in the 21st century.

Racism and institutional bias remain a profound existential threat around-the-world manifesting not only in social and economic disparities but also deeply impacting the healthcare delivery system (Yearby et al., 2022, Feagin et al., 2014). Systemic racism results in unequal access to healthcare, poorer health outcomes, and higher mortality rates for marginalized communities, particularly in people of color in both HICs and LMICs. Discrimination and institutional bias within the global healthcare system has led to mistrust among minority populations world-wide, deterring them from seeking necessary medical care. This mistrust is compounded by historical injustices and ongoing disparities in treatment and diagnosis. Addressing racism in global healthcare delivery requires comprehensive policy changes, cultural competency training for healthcare providers, and a commitment to equity and justice to ensure that all individuals receive fair and adequate medical attention.

Natural disasters pose a significant threat to the global healthcare delivery system by overwhelming medical infrastructure, disrupting supply chains, and displacing populations (Ferguson et al., 2024). Hurricanes, wildfires, earthquakes, and floods can destroy hospitals and clinics, leading to a sudden loss of essential healthcare services. In the aftermath of such disasters, there is often an increase in health emergencies, including injuries, infectious diseases, and mental health crises, which further strain the already stretched resources. Effective emergency (i.e., disaster) preparedness and rapid response plans, including robust infrastructure, efficient emergency medical services, and resilient healthcare systems, are essential to mitigate the impact of natural disasters on global healthcare delivery. Additionally, ensuring that marginalized and vulnerable populations have access to safe, timely, efficacious, effective equitable, and person-centered (IOM, 2001) medical care during and after disasters is crucial for the ever-widening gap of health disparities.

Emerging infectious diseases (EIDs), such as COVID-19 and Ebola, pose a grave threat to the global healthcare delivery system by rapidly increasing the demand for medical services and exposing weaknesses in multi-level public health preparedness and response (Palagyi et al., 2019; Baker et al., 2022) These diseases can lead to widespread illness, complex chronic conditions, and death, placing immense pressure on local hospitals, healthcare workers, and medical supplies. The COVID-19 pandemic highlighted the critical need for robust multi-level early disease surveillance systems, rapid and coordinated response capabilities, and effective communication strategies to manage such crises. Additionally, it underscored the importance of global cooperation and investment in research and development for new evidence-based vaccines and treatments. Strengthening a global and chronically-underfunded public health infrastructure, raising the bar on efficient and effective infection control principles and practices, and ensuring equitable access to healthcare service delivery are vital to mitigate the impact of EIDs on the global healthcare system.

Economic inequality and inequity profoundly affect the global healthcare delivery system by creating barriers to access and perpetuating health disparities (Dickman et al., 2017; Horowitz et al., 2020). People of lower socioeconomic status frequently encounter challenges such as insufficient health insurance coverage, elevated out-of-pocket expenses, limited access to quality healthcare, and the burden of mounting medical debt. These socio-economic barriers result in delayed diagnoses and treatment of acute and chronic disease and poorer overall health outcomes for disadvantaged populations. Furthermore, healthcare providers serving LMICs may face significant financial constraints, limiting their ability to offer comprehensive and essential health services. Addressing economic inequality in healthcare requires public policy interventions to expand insurance coverage for all, reduce costs, close health disparities, and invest in underserved areas. Ensuring that all individuals, regardless of their socio-economic status, have access to high-quality healthcare is essential for promoting equity and improving both individual and community health outcomes.

Finally, Anthropocene climate change—driven by human activities—has a profound impact on global healthcare delivery. It intensifies environmental health challenges and heightens both the frequency and severity of natural disasters, further straining healthcare systems worldwide (Ebi et al., 2021; Huggel et al., 2022; Seervai et al., 2022; Eckelman et al., 2020). Rising ambient temperatures, air, water, and soil pollution, and extreme weather events contribute to a range of health problems, including respiratory diseases, heat-related illnesses, and vector-borne diseases. Vulnerable populations, such as the elderly, children, and those with chronic conditions, are particularly at risk. The global healthcare delivery system must adapt to these changes by developing resilience, improving emergency preparedness and rapid response, and integrating climate considerations into multi-level medical care and public health planning. Additionally, addressing the root causes of Anthropocene climate change through sustainable health, safety, and welfare (HSW) principles, practices, and policies is crucial to mitigate its long-term effects on individual and community health and healthcare service delivery. Taking proactive measures to address the impacts of human-induced climate change is essential to safeguard individuals, preserve the communities where they live, work, and play, and ensure the resilience and sustainability of global healthcare systems in the face of environmental challenges.

1.2 Urgent Appeal: The Global Healthcare Delivery System's Role in Addressing Existential Challenges.

The global healthcare delivery system is at a critical juncture, facing urgent challenges that threaten its ability to deliver integrated, holistic, comprehensive, effective, equitable, and sustainable essential health services. Among these challenges are escalating healthcare costs, workforce fatigue and burnout, informatics and cybersecurity vulnerabilities, political and regulatory instability, institutional bias and systemic racism, economic inequality, human-induced climate change and greenhouse gas (GHG) emissions, natural disasters, pandemics, and the growing burden of chronic diseases. Addressing these existential threats demands immediate, coordinated efforts worldwide through traditional medical care (e.g., advanced primary care services embedded in a patient-centered medical home model), essential public health services (e.g., health screenings, immunizations, environmental health), and community-based organizations' services (e.g., community health workers and peer health system navigators). To safeguard its future, the global healthcare delivery system must prioritize resilience, innovation, and equity in shaping its policies and practices (Witter et al., 2023; Biddle et al., 2020; Nzinga et al., 2021).

The global healthcare delivery system must prioritize building resilience to the impacts of human-induced climate change (Lugten et al., 2022). This effort includes preparing for extreme weather events—such as hurricanes, wildfires, and heatwaves—that can disrupt essential health services and worsen acute and chronic health conditions. Hospitals and healthcare facilities should, whenever feasible, embrace environmentally sustainable practices, including reducing carbon footprints, enhancing energy efficiency, and transitioning to renewable energy sources, if available. Moreover, healthcare providers should incorporate climate change considerations into traditional medical care and public health planning, addressing the effects of environmental factors on health outcomes with proactive, integrated, and person-centered systems of health (e.g., disease prevention, health promotion, health protection, disease surveillance with population health management, and emergency preparedness, resilience, and rapid response).

The COVID-19 global pandemic highlighted the critical need for robust world-wide public health multi-level infrastructure, preparedness, and response (Metzl et al., 2020; Niles, 2023). Strengthening the global healthcare delivery system's ability to respond to future pandemics requires enhanced surveillance, rapid response capabilities, and effective communication strategies. Investing in research and development for vaccines, therapeutics, and diagnostic tools is essential for staying ahead of emerging infectious disease (EIDs). Moreover, building a flexible and highly-trained healthcare workforce capable of scaling up in times of crisis, along with ensuring adequate supplies of personal protective equipment (PPE) and other medical supplies and resources, is critical for maintaining essential healthcare services during pandemics.

Chronic diseases, including diabetes, hypertension, heart disease, respiratory dysfunction, cancer, neurodegenerative disorders, and obesity, present significant, enduring challenges to the global healthcare system due to their rising incidence, prevalence, and associated financial burden (Crawford et al., 2023). Shifting the continuum of care towards proactive measures and promoting healthy lifestyles and behaviors before the irreversible onset of symptoms is imperative. Healthcare providers should prioritize early detection of disease through evidence-based screening practices (e.g., United States Preventive Services Task Force recommendations), patient education initiatives, and enhancing health literacy. Community-based interventions led by local organizations that foster healthy behaviors are equally vital. Moreover, leveraging technology—such as telehealth and remote patient monitoring—can strengthen chronic disease management while alleviating the strain on healthcare facilities already operating under immense pressure. Policymakers and regulatory bodies must address the upstream health-related social needs (HRSNs) and other determinants of health, including access to nutritious food, safe housing, adequate transportation options, and opportunities for physical activity, to mitigate the occurrence of chronic diseases and foster improvements in both individual and population health.

Elevating representation, fairness and belonging must be a top priority in the global healthcare system's efforts to combat economic inequality, systemic inequity, institutional bias, and racism (Stanford, 2020). Marginalized communities bear a disproportionate burden from these challenges, underscoring the urgency of ensuring equitable access to essential healthcare services. To achieve this, it is critical to address systemic barriers, such as racial and socioeconomic disparities and discriminatory practices like redlining. Every individual and community—regardless of background—must have access to high-quality healthcare where they live, work, and play.

Community engagement and partnerships with local organizations are vital to developing culturally sensitive interventions that effectively serve underserved populations. Additionally, robust data collection and analysis should be leveraged to identify health inequities and inform targeted strategies that promote health equity and health literacy.

The global healthcare delivery system must act decisively to address the many existential threats described (Laprise, 2023). By adopting a proactive, integrated, comprehensive and holistic approach to providing a full range of healthcare services, the global healthcare delivery system can better protect the health, wellness, and well-being of individuals, families and communities creating a more environmentally sustainable and equitable future for all. No one's quality or length of life should be dictated by their zip code on Earth.

1.3 Healthcare Stewardship: A Framework for Responding to Global Existential Threats.

Healthcare stewardship represents a vital response to the global healthcare system's urgent need to confront existential threats (Brinkerhoff et al., 2019). By focusing on the responsible management of finite resources, proactive care, and long-term environmental sustainability, stewardship ensures that healthcare systems worldwide remain resilient, equitable, and effective. As the global healthcare landscape contends with challenges like anthropogenic climate change, pandemics, and the growing prevalence of chronic diseases, the principles and practices of stewardship provide a robust framework to address these pressing issues.

Healthcare stewardship entails the prudent management of limited resources to maximize health outcomes while minimizing waste and inefficiency. This approach calls for a transition to empirically-driven, value-based care that emphasizes high-quality, evidence-based clinical practices delivered at the lowest total cost, ultimately leading to improved patient outcomes. By prioritizing proactive strategies and early interventions at both individual and community levels, global healthcare systems can mitigate the burden of chronic diseases and prevent expensive hospitalizations.

In addition, stewardship extends to adopting sustainable environmental practices, such as energy-efficient healthcare facilities, reducing greenhouse gas emissions, and promoting environmental health. These efforts are critical to lessening the impact of anthropogenic climate change on public health.

In order to achieve the goals and objectives of healthcare stewardship described above, The World Health Organization's (WHO) primary health care (PHC) model serves as a best-in-class solution (WHO, 2025). The WHO's PHC model is a comprehensive, whole-of-society approach designed to strengthen global healthcare delivery systems and bring essential health services closer to communities. It is built on three core components:

Integrated Health Services: PHC emphasizes meeting people's health needs throughout their lives, from health promotion and disease prevention to treatment, rehabilitation, and palliative care. It ensures that care is people-centered, respecting individual needs, preferences, and values.

Addressing Broader Upstream Determinants of Health: The model recognizes the importance of interdisciplinary and multisectoral public policies and actions to tackle social, economic, and environmental factors that influence health outcomes. This includes improving water and sanitation, addressing noncommunicable diseases, and preparing for health emergencies.

Empowering Communities: PHC focuses on enabling individuals, families, and communities to take charge of their health. This involves community engagement, culturally appropriate interventions, and platforms for consultation to identify health priorities.

The PHC model is regarded as the most inclusive, equitable, and cost-effective way to achieve universal health coverage (UHC) and the health-related Sustainable Development Goals (SDGs). It also enhances the resilience of health systems, enabling them to respond effectively to shocks and crises.

The COVID-19 global pandemic underscored the need for rapid response capabilities, including the ability to quickly develop and distribute vaccines and therapeutics (Wong et al., 2022; NASEM, 2021). By fostering a culture of emergency preparedness, rapid response, and continuous quality improvement, global healthcare delivery systems can better respond to pandemics and other global emerging threats. Proactive measures within the WHO PHC model also includes integrating advance medical and information technology with computational science and big data analytics to enhance disease surveillance, predict outbreaks, and tailor interventions to the specific needs of individuals and their communities (WHO, 2025).

Equity is a cornerstone of healthcare stewardship. To build a resilient and just global healthcare delivery system, it is crucial to ensure that all individuals have access to high-quality care, irrespective of their socioeconomic status, race, or geographic location. Achieving this requires addressing upstream health-related social needs (HRSNs) and dismantling systemic barriers that perpetuate health disparities and health illiteracy. Advocacy across all socio-ecological levels worldwide must champion evidence-based public policies that advance health equity and health literacy, while fostering collaboration with community-based organizations to create culturally sensitive and accessible integrated healthcare systems. By placing equity at the center of healthcare stewardship, vulnerable populations can be safeguarded, and the benefits of healthcare innovations can be equitably distributed, ensuring broad and lasting impact.

In conclusion, healthcare stewardship provides a strategic and ethical approach for responding to the existential threats facing the global healthcare delivery system. By emphasizing resource management, proactive care, and equity, stewardship ensures that healthcare systems are resilient, sustainable, and capable of delivering high-quality and cost-effective healthcare goods and services to all individuals in the right amount and at the right place. Embracing the principles of stewardship is crucial for addressing the complex and interconnected challenges of man-made climate change, pandemics, and chronic diseases, ultimately leading to healthier and more equitable communities.

1.4 Examples of Healthcare Stewardship

Examples of healthcare stewardship span various aspects of global healthcare delivery, policy-making, and resource management (Nasiri et al., 2019; Shams et al., 2021, Duplechan, 2022).

1. *Resource Allocation*
   a) *Prioritizing Vaccinations*: During flu seasons or pandemics, health authorities prioritize vaccination for high-risk groups, such as the elderly, healthcare workers, and individuals with chronic conditions. This ensures that limited vaccine supplies are used where they can have the greatest impact.
   b) *Essential Medicines List*: The World Health Organization (WHO) maintains a list of essential medicines that should be available to all healthcare systems around-the-world. This helps countries prioritize their available resources and ensure access to the most critical medications.

2. *Evidence-Based Practice*
   a) *Clinical Guidelines*: Developing and adhering to empirically-derived clinical practice guidelines for treating common conditions ensures that healthcare providers use the most effective and efficient treatment methods. This reduces variations in medical care and improves patient outcomes.

   b) *Proactive Measures' Initiatives*: Evidence-based programs that promote disease prevention, such as routine screenings for cancer or heart disease, help detect conditions early, reducing the need for more extensive and costly treatments further downstream.

3. *Health System Management*
   a) *Integrated Care Models*: Implementing integrated care models that coordinate services across primary, secondary, and tertiary care settings can reduce redundancies, improve patient outcomes, and lower healthcare costs.
   b) *Electronic Health Records (EHRs)*: Utilizing EHRs to streamline information sharing among healthcare providers can enhance care coordination, reduce medical errors, and improve the efficiency of healthcare delivery.

4. *Environmental Stewardship*
   a) *Sustainable Practices in Healthcare Facilities*: Hospitals and clinics adopting environmentally sustainable practices, such as reducing waste, recycling medical materials, and using energy-efficient technologies, contribute to the overall sustainability of the healthcare system.
   b) *Green Procurement Policies*: Healthcare organizations that prioritize the procurement of environmentally friendly products and services help reduce their ecological footprint.

5. *Ethical and Equitable Care*
   a) *Community Health Programs*: Initiatives that provide essential health services to underserved and marginalized populations, such as mobile health clinics and community health worker programs, ensure that all individuals have access to traditional primary care and essential public health services.
   b) *Health Equity Policies*: Developing and implementing public policies that address health-related social needs (HRSNs),

such as stable and safe housing, food security, transportation, education, and income, helps create a more equitable healthcare delivery system.

6. *Policy and Governance*
   a) *Regulatory Oversight*: Governments, regulatory agencies, and public health authorities provide oversight to ensure that healthcare providers adhere to standards of care, maintain patient safety, and use resources efficiently.
   b) *Public Health Campaigns*: Campaigns to raise awareness about healthy lifestyles, vaccinations, and disease prevention are crucial in managing essential public health services and resources and preventing outbreaks of diseases.

These examples demonstrate how healthcare stewardship around-the-world embodies a holistic and integrated strategy for resource management, enhancing healthcare quality and safety, reducing total cost of care, and ensuring equitable access to healthcare services.

1.5 References

1. Baker, R. E., Mahmud, A. S., Miller, I. F., et al. (2022). Infectious disease in an era of global change. *Nature Reviews Microbiology, 20*(4), 193-205.
2. Berman, E. P. (2022). Thinking like an economist: How efficiency replaced equality in US public policy. Princeton, NJ: Princeton University Press.
3. Biddle, L., Wahedi, K., Bozorgmehr, K. (2020). Health system resilience: a literature review of empirical research. *Health policy and planning 35*(8) 1084-1109.
4. Brinkerhoff, D.W., Cross, H.E., Sharma S., et al. (2019) Stewardship and health systems strengthening: An overview. *Public Admin Dev* 39: 4–10.

5. Chan, J. C., Lim, L. L., Wareham, N. J., et al. (2020). The Lancet Commission on diabetes: using data to transform diabetes care and patient lives. *The Lancet 396*(10267): 2019-2082.
6. Crawford, A. M., Shiferaw, A. A., Ntambwe, P., et al. (2023). Global critical care: a call to action. *Critical Care*, *27*(1), 28.
7. de Vries, N., Boone, A., Godderis, et al. (2023). The Race to Retain Healthcare Workers: A Systematic Review on Factors that Impact Retention of Nurses and Physicians in Hospitals. *Inquiry. 60*: 469580231159318.
8. Dickman, S. L., Himmelstein, D. U., Woolhandler, S. (2017). Inequality and the health-care system in the USA. *The Lancet*, *389*(10077), 1431-1441.
9. Duplechan, L. (2022) How health care organizations can be stewardship leaders. *AMA J Ethics* 24(10): e1013-e1021.
10. Ebi, K. L., Vanos, J., Baldwin, J. W., et al. (2021). Extreme weather and climate change: population health and health system implications. *Annual review of public health*, *42*(1), 293-315.
11. Eckelman, M., et al. (2020) Health Care Pollution and Public Health Damage in the United States: An Update. *Health Affairs.* 39(12): 2071–79.
12. Feagin, J., Bennefield, Z. (2014). Systemic racism and US health care. *Social science & medicine. 103*: 7-14.
13. Ferguson, A., Ward, K., Parke, R. (2024). What is known about resilient healthcare systems in the context of natural disasters? A scoping review. *Collegian.*
14. Horowitz, J. M., Igielnik, R., Kochhar, R. (2020). Most Americans say there is too much economic inequality in the US, but fewer than half call it a top priority. *Pew Research Center*, *9.*
15. Huggel, C., Bouwer, L.M., Juhola, S. *et al.* (2022) The existential risk space of climate change. *Climatic Change* 174, 8.
16. Institute of Medicine (IOM) Committee on Quality of Health Care in America. (2001). *Crossing the Quality Chasm: A New Health System for the 21st Century.* National Academies Press, Washington, DC.

17. Lal, A., Erondu, N. A., Heymann, D. L., et al. (2021). Fragmented health systems in COVID-19: rectifying the misalignment between global health security and universal health coverage. *The Lancet, 397*(10268), 61-67.
18. Laprise, C. (2023). It's time to take a sustainable approach to health care in the face of the challenges of the 21st century. *One Health* 16: 100510.
19. Li, W., Chai, Y., Khan, F., et al. (2021). A comprehensive survey on machine learning-based big data analytics for IoT-enabled smart healthcare system. *Mobile networks and applications, 26*, 234-252.
20. Lopes, L., Montero, A., Presiado, M., et al. (2024) Americans' Challenges with Health Care Costs. *KFF Brief.*
21. Lugten, E., & Hariharan, N. (2022). Strengthening Health Systems for Climate Adaptation and Health Security: Key Considerations for Policy and Programming. *Health security 20*(5): 435–439.
22. Metzl, J. M., Maybank, A., De Maio, F. (2020). Responding to the COVID-19 pandemic: the need for a structurally competent health care system. *JAMA 324*(3): 231-232.
23. Nasiri, T., Takian, A., & Yazdani, S. (2019) Stewardship in Health, Designing a Multi-Layer Meta Model: A Review Article. *Iranian journal of public health 48*(4): 579–592.
24. National Academies of Sciences, Engineering, and Medicine. (2021) Learning from Rapid Response, Innovation, and Adaptation to the COVID-19 Crisis: Proceedings of a Workshop in Brief. Washington, DC: The National Academies Press.
25. Niles, N. J. (2023). *Basics of the US health care system.* Burlington, MA: Jones & Bartlett Learning.
26. Nzinga, J., Boga, M., Kagwanja, N., et al. (2021). An innovative leadership development initiative to support building everyday resilience in health systems. *Health policy and planning 36*(7): 1023-1035.
27. Okafor, C.M., Kolade, A., Onunka, T., et al. (2023) Mitigating cybersecurity risks in U.S. healthcare sector. IJRSI. 10(9): 177-194.

28. Palagyi, A., Marais, B., Abimbola, S., et al. (2019) Health system preparedness for emerging infectious diseases: A synthesis of the literature. *Global Public Health* 14(12): 1847 -1868.
29. Seervai, S., Gustafsson, L., Abrams, M.K. (2022) How the U.S. Health Care System Contributes to Climate Change *Commonwealth Fund.*
30. Shams, L., Sari, A. A., Yazdani, S., Nasiri, T. (2021). Model for Value-based Policy-making in Health Systems. *International journal of preventive medicine 12*: 13.
31. Stanford F. C. (2020). The Importance of Diversity and Inclusion in the Healthcare Workforce. *Journal of the National Medical Association, 112*(3), 247–249.
32. Thomas, S., Sagan, A., Larkin, J., et al. (2020). Strengthening health systems resilience: key concepts and strategies. Copenhagen, Denmark: European Observatory on Health Systems and Policies.
33. Vodovotz, Y., Barnard, N., Hu, F. B., et al. (2020). Prioritized research for the prevention, treatment, and reversal of chronic disease: recommendations from the lifestyle medicine research summit. *Frontiers in medicine 7*: 585744.
34. Witter, S., Thomas, S., Topp, S. M., et al. (2023). Health system resilience: a critical review and reconceptualisation. *The Lancet Global Health, 11*(9), e1454-e1458.
35. Wong, L., Qureshi, K., Glauberman, G., et al. (2022). Rapid Response: The Development of Just-in-Time Education for Nursing Clinicians and Students. *Hawai'i journal of health & social welfare,* 81(1): 21–24.
36. World Health Organization (WHO). (2025). *Primary Health Care Fact Sheet.* World Health Organization.
37. Yearby, R., Clark, B., Figueroa, J. F. (2022) Structural Racism in Historical and Modern U. S. Health Care Policy. Health Affairs 41(2): 187-194.

2.0

Healthcare Stewardship in a Modern-Day Society.

2.1 The Roots of Stewardship.

The concept of stewardship originates from biblical times. According to Saltman and Ferroussier-Davis (2000), the Old Testament presents the following references to stewardship:

1. In the Book of Genesis, men and women were created in the image of God, given a privileged place among creatures, and commanded by God to exercise stewardship over the earth.
2. Also in Genesis, Jacob's son, Joseph, when sold into slavery by his brothers becomes Potiphar's then Pharaoh's steward, and exhibits the following qualities of stewardship:
    a) He becomes a selfless servant of his masters.
    b) He manages his master's assets without owning any of them.
    c) He anticipates the future trends and devises grand plans for the use of his master's assets.

In the New Testament, Jesus tells the parable of the talents, whereby a master divides his goods between his three servants. The story emphasizes that when one is entrusted with something of value by others, one has an obligation ethically and morally to improve the value of what they are managing. The notion of stewardship as an ethically driven responsibility for protecting one's precious and limited resources lies at the heart of both the Christian and Jewish faiths.

Saltman and Ferroussier-Davis (2000) also describe the very first reference to healthcare stewardship based on religious teachings. The Islamic faith and the institution of Hisba organized public administrative functions in both the moral/normative and administrative/technical dimensions. The head of Al Hisba, the first Muhtasib, was appointed in Medina in the ninth century. The functions of the Muhtasib in precolonial Arab societies included the regulation of medical practice and pharmaceuticals and overseeing the requirements regarding the equitable provision of services and the public interest.

Religion also serves as the basis for the environmental and ecosystem stewardship that has emerged in Europe and North America within the past decade (Saltman and Ferroussier-Davis 2000). Originally, it started out as accountability to God in the Judeo-Christian doctrine. Responsibility to God has been replaced within the environmental and ecosystem movement with intergenerational responsibility. Saltman and Ferroussier-Davis (2000) interpret the lack of control over the degradation of the environment over centuries as humans' misinterpretation of God's mandate to be earth's steward as a license to dominate, exploit, and destroy the environment. Others argue that damage to the environment has occurred because of people disregarding God's recommendations to appropriately manage the earth's valuable resources.

2.2 Defining Healthcare Stewardship

Healthcare stewardship refers to the responsible management and utilization of limited healthcare resources (i.e., goods and services) to ensure the sustainability, efficiency, and effectiveness of healthcare delivery. It involves a broad range of activities aimed at optimizing health outcomes, improving the quality and safety of care, and ensuring equitable access to healthcare services. The concept emphasizes the need for global healthcare delivery systems to operate in a manner that balances resource allocation, ethical considerations, and long-term planning to meet the current and future health needs of the world's population (Brinkerhoff et al., 2019a).

At its essence, healthcare stewardship involves making thoughtful decisions to achieve optimal health outcomes while reducing waste and inefficiencies. This entails the careful management of financial resources, medical supplies, and human capital, alongside the adoption of policies and practices that prioritize the well-being of individuals and communities. Successful stewardship depends on collaboration among diverse stakeholders—government agencies, healthcare providers, patients, and the broader community—to ensure resources are utilized wisely and the healthcare system remains adaptable, resilient, and responsive to evolving demands (Saltman et al., 2000; Travis et al., 2002, Greer 2018).

Healthcare stewardship encompasses the ethical responsibility to deliver high-quality care while addressing health disparities and ensuring equitable access to essential health services for all individuals without causing undo financial hardship. It involves prioritizing clinical interventions based on empirical effectiveness and cost-efficiency, advancing proactive measures, primary health care as the cornerstone of healthcare services delivery, and fostering a culture of transparency, accountability, integrity, and continuous quality improvement within healthcare organizations. By responsibly managing limited healthcare resources, stewardship works to build trust and confidence in the global healthcare system, ultimately driving better health outcomes worldwide (Brinkerhoff et al., 2014; Stanford, 2023; WHO, 2020).

Moreover, healthcare stewardship has become increasingly relevant amidst global health challenges, including pandemics, aging populations, and the rising prevalence of chronic diseases. As global healthcare delivery systems confront these complex issues, the principles of stewardship provide a foundation for developing empirically-driven innovative solutions, adopting evidence-based clinical practice guidelines, and aligning health-related public policies with broader societal objectives. By emphasizing sustainable and equitable healthcare delivery, stewardship ensures that healthcare systems are equipped to tackle current challenges and maintain resilience to adapt to future demands (Brinkerhoff et al., 2019b).

2.3 The Core Principles of Healthcare Stewardship

Healthcare stewardship is underpinned by several core principles that guide the responsible management, allocation, and utilization of healthcare resources. These principles aim to ensure that global healthcare delivery systems operate efficiently, ethically, and sustainably to meet the needs of all individuals and populations (Kapoor et al., 2014).

A core principle of healthcare stewardship is the efficient and effective utilization of limited healthcare resources. This entails strategically optimizing the allocation of financial, human, and material resources to achieve the best possible health outcomes. Efficient global healthcare systems emphasize minimizing waste, eliminating redundancies, and streamlining processes to ensure resources are directed where they can create the greatest impact. Key approaches include adopting value-based care models, implementing empirically-driven evidence-based clinical practices, and fostering proactive, predictive, and personalized health systems. Additionally, leveraging data analytics plays a pivotal role in guiding shared decision-making, community-oriented primary health care (COPHC), and enhancing overall effectiveness of individual- and population-based healthcare service delivery (Abrams et al., 2021).

Equity and accessibility are both foundational principles to healthcare stewardship (White-Williams et al., 2022). Ensuring that all individuals have fair access to healthcare services, regardless of their socioeconomic status, geographic location, or other upstream determinants of health, is essential for achieving health equity. This principle emphasizes the need to address health disparities in the equitable access to essential healthcare services by removing obstacles and barriers and providing targeted support to underserved and vulnerable populations. Policies and practices that promote inclusive healthcare, such as expanding insurance coverage, enhancing primary health care services, and addressing upstream health-related social needs (HRSNs), are critical to the collaborative and coordinated implementation of this principle.

Accountability and transparency are essential for fostering trust and maintaining public confidence in the global healthcare delivery system (Meagher et al., 2022). Healthcare stewardship demands that all stakeholders—including healthcare providers, organizations, and policymakers—act with transparency and integrity in their decisions and actions. This entails establishing clear norms and standards, tracking and reporting key performance indicators (KPIs), and ensuring the ethical and appropriate allocation of healthcare resources. Transparent practices in decision-making, resource distribution, and reporting outcomes not only promote accountability but also empower stakeholders to assess the fairness and effectiveness of the healthcare delivery systems that provide their care.

Sustainability is a key principle of healthcare stewardship, focusing on the long-term capability of the global healthcare delivery system (Hu et al., 2022). This involves balancing the immediate needs of the population with the preservation of healthcare resources for future generations. Sustainable healthcare practices include investing in proactive and primary health care, promoting healthy lifestyles and behaviors, and adopting environmentally-friendly nature-based solutions within healthcare facilities. By prioritizing sustainability, individual healthcare delivery systems around-the-world can ensure that they develop and remain resilient and capable of mitigating and adapting to future challenges, such as demographic shifts, information and technological advancements, and emerging existential health threats.

Ethical considerations form the cornerstone of responsible decision-making in healthcare stewardship, shaping the equitable distribution of resources and upholding respect for patient rights and autonomy. (Guidolin et al., 2022). This principle involves stakeholders making decisions, both clinical and administrative, based on ethical frameworks that consider the needs, preferences, and values of individuals and communities. Ethical healthcare stewardship includes respecting patient confidentiality, obtaining informed consent prior to every intervention, and ensuring that healthcare services are provided in a manner that respects the dignity and rights of all patients. It also involves addressing ethical dilemmas, such as resource allocation during crises, through transparent and inclusive decision-making processes.

Effective healthcare stewardship relies on collaboration and partnerships among diverse stakeholders, including government agencies, healthcare providers, patients, and communities (Robu et al., 2020). This principle underscores the significance of collective action to achieve shared objectives, social justice, and positive health outcomes. Cooperative efforts facilitate resource sharing, the dissemination of empirically-driven evidence-based best clinical practices, and foster innovation and creativity in healthcare service delivery. Furthermore, alliances with community-based organizations, academic medical centers, and international regulatory bodies play a critical role in addressing intricate global healthcare system challenges and ensuring a cohesive and comprehensive approach to healthcare stewardship.

2.4 What Are The Key Forces Driving Healthcare Stewardship?

Previously defined, *Healthcare Stewardship* refers to the responsible management, allocation, and utilization of limited healthcare resources to achieve optimal health outcomes for individuals and populations. It involves the careful planning, regulation, and oversight of individual healthcare systems by its stakeholders to ensure efficiency, equity, quality, and environmental sustainability. Healthcare stewardship is driven by several foundational forces that shape its principles, policies, practices, and overall approach to empirically-driven healthcare management. These forces are generally categorized into economic, social, political, and technological dimensions. (He et al., 2022; Wilson et al., 2020; Kaiser et al., 2022; Tello et al., 2020; Scott et al., 2021; Hubbard et al., 2022; Dion et al., 2024; MacNeill et al., 2021; Fatemi et al., 2022; Shaw et al., 2021).

1. *Economic Factors.*
    a) *Cost Containment*: Rising global healthcare costs are a major concern for governments, insurers, and patients. Effective stewardship aims to control the total costs of care without compromising patient quality and safety by promoting efficient use of limited resources, reducing fraud, waste, and abuse, and implementing cost-effective interventions.
    b) *Resource Allocation*: Equitably distributing limited healthcare resources to address the diverse needs, preferences, and values of various population groups is essential. This entails prioritizing primary health care (PHC) services across both individual and population socio-ecological levels of organization and focusing on the most cost-effective interventions that optimize positive health outcomes for patients.
    c) *Value-Based Care*: Emphasizing value over volume, this approach focuses on achieving better health outcomes per dollar spent, incentivizing providers to deliver high-quality,

safe, timely, effective, efficacious, and equitable patient-centered care (IOM, 2001).

2. *Social Factors.*
    a) *Health Equity*: closing the gaps in health disparities for access and navigation to appropriate healthcare services is a key goal of healthcare stewardship. This involves ensuring that vulnerable and marginalized populations receive evidence-based best-in-class clinical practices in-person or by telehealth with remote patient monitoring and that all upstream determinants of health are addressed.
    b) *Population Health Management*: Focusing on the health outcomes of entire populations rather than just individual patients, stewardship efforts aim to improve community-health status through health prevention, health promotion, health protection, disease surveillance, and emergency preparedness, resilience and rapid response coordinated by advanced primary care services, essential public health services, and community health worker navigation assistance from community-based organizations for acute and chronic conditions.
    c) *People-Centered Care*: Engaging people in their own care towards self-efficacy and self-actualization and ensuring that healthcare services are responsive to their cultural needs, preferences and values is essential for effective stewardship (Khatri et al., 2023).

3. *Political Factors.*
    a) *Regulation and Oversight*: Governments play a critical role in regulating healthcare delivery systems, setting regulatory standards for quality, safety, and ethics, and ensuring compliance through monitoring and enforcement.
    b) *Policy and Legislation*: Healthy public policies and laws shape the structure and functioning of global healthcare

systems. Effective stewardship requires public policy-making that is informed by empirical evidence and aimed at achieving equitable and efficient healthcare delivery.

c) *Public Accountability*: Ensuring transparency and accountability in healthcare decisions and spending is foundational to healthcare stewardship. This includes involving stakeholders in decision-making processes and regularly reporting on healthcare's key performance indicators and outcomes that matter the most to patients. Use of real-time dashboards and annual scorecards are examples of tools that transparently present healthcare outcomes in different formats for different purposes for different populations.

4. *Technological Factors.*
   a) *Health Information Technology*: The use of electronic health records (EHRs), telemedicine, and health information exchanges (HIEs) enhances the efficiency, quality, and coordination of care. Effective stewardship involves leveraging these technologies to capitalize on the interoperability of current information systems to improve health outcomes by using encounter/claims-based data with EHR-derived clinical information and streamline healthcare processes.
   b) *Innovation and Research*: Advancing biomedical research, fostering learning health ecosystems, and promoting innovation in healthcare services and cutting-edge medical technologies enable the early detection of disease through specific biomarkers, even before symptoms emerge. These efforts also enhance treatment options and ensure more cost-effective healthcare delivery. Effective stewardship plays a vital role in supporting and regulating these advancements to maximize their benefits for both individuals and society as a whole.
   c) *Data Analytics*: Leveraging big data, computational science, systems medicine, and advanced analytics is integral to mod-

ern healthcare stewardship. These tools enhance provider and patient decision-making, monitor health outcomes, and pinpoint opportunities to reduce health disparities.

5. *Ethical and Moral Factors.*
   a) *Ethical Decision-Making*: Healthcare decisions must be guided by ethics, cultural humility, and empathy, with a focus on the health, wellness, and well-being of patients. Equally vital are principles of reasonableness and social justice, which consider the needs of individuals and the broader community. This approach requires making challenging decisions about resource allocation and prioritizing interventions, always grounded in well-established ethical principles.
   b) *Patient Rights*: Safeguarding patient rights—such as informed consent, privacy, and autonomy—is a cornerstone of ethical healthcare. Effective stewardship goes beyond merely protecting these rights; it actively fosters an environment where patients are empowered to make decisions about their care, fully understanding their options and the potential consequences. Moreover, stewardship integrates these principles into a broader commitment to achieving optimal health outcomes, ensuring that ethical practices and patient well-being remain central to every decision within the healthcare delivery system.

6. *Environmental Factors.*
   a) *Sustainable Practices*: Promoting environmentally sustainable practices within healthcare delivery systems and hospitals is an essential aspect of modern healthcare stewardship. This includes initiatives such as reducing waste through improved recycling programs, minimizing single-use plastics, conserving energy with efficient lighting and HVAC systems, and adopting renewable energy sources. Furthermore, incorporating eco-friendly materials in construction, medical

supplies, and daily operations can significantly reduce the healthcare sector's environmental footprint. By prioritizing sustainability, healthcare institutions not only contribute to global environmental goals but also foster healthier communities by mitigating the impacts of man-made climate change and air, water, and soil pollution. Responsible stewardship integrates these efforts into a broader commitment to ethical and equitable healthcare services delivery.

b) *Anthropocene Climate Change and Impact on Health*: Tackling the health impacts of anthropogenic (i.e., human-induced) climate change and equipping healthcare systems to address climate-related health challenges are integral elements of contemporary healthcare stewardship. This includes understanding and mitigating the direct effects of climate change, such as the rise in heat-related illnesses, the spread of vector-borne diseases, and the increased frequency of extreme weather events that disrupt healthcare access. It also involves strengthening healthcare infrastructure to adapt to these threats, ensuring resilience in supply chains, emergency preparedness with rapid response to pandemics and natural disasters, and resource allocation during climate-related crises. Furthermore, fostering collaboration between healthcare organizations, governments, and global entities to develop public policies and interventions that prioritize sustainability and climate resilience is critical. By proactively addressing these challenges, healthcare stewardship protects public health while contributing to broader efforts to combat the existential threat posed by climate change.

c) *Nature-based Solutions*: Biodiversity conservation encompasses a broad and sustainable approach by society to protect and preserve ecosystems' services while addressing critical social factors. These include promoting socio-economic development and fostering effective leadership and governance to ensure equitable and impactful stewardship

> of natural resources. Sustainable biodiversity management provides vital ecosystem services that enhance the urban biosphere, such as regulating micro-climates to counteract urban heat effects, preventing floods through improved water management, and ensuring water treatment for cleaner and safer resources. Additionally, it facilitates food provision, supports pollination, and contributes to overall ecosystem balance. By integrating these practices, society can foster healthier environments while advancing social and economic resilience in harmony with nature.

Healthcare stewardship is influenced by the dynamic interplay of economic, social, political, technological, ethical, and environmental factors. Its overarching goal is to achieve positive health outcomes, optimize wellness, enhance overall well-being, and ensure equity and accessibility in healthcare service delivery. Effective stewardship also seeks to control healthcare costs while championing environmentally sustainable and ethical practices. By tackling these complex challenges, global healthcare delivery systems can better serve their communities and special populations, adapting to the demands and complexities of an ever-evolving global healthcare delivery system landscape.

2.5 Unlocking Opportunities to Enhance Public Health Policy Through Healthcare Stewardship.

The positive dimension of stewardship underscores its immense potential to enhance healthy public policy outcomes, as highlighted by Saltman et al. (2000), Redvers (2021), Woolhandler et al. (2021), and Bak-Coleman et al. (2021). Central to this debate is the challenge of anchoring the healthy public policy process within traditional principal-agent dynamics, which prioritize creating economically stable healthcare systems. While economic viability remains a critical goal, stewardship offers an opportunity to transcend purely financial considerations. By emphasizing normative, ethically driven expectations, healthcare stewardship can foster a global healthcare delivery system that is not only cost-efficient but also guided by principles of equity, compassion, and the common good. This approach integrates ethical leadership and collaborative governance to create a global healthcare delivery system that reflects broader societal values and contributes to sustainable, holistic wellness and well-being.

Stewardship should act as a guiding principle in the global healthcare delivery system, ensuring that power is exercised to prioritize the common good over purely economic interests. By placing the collective well-being at the forefront, stewardship has the potential to reinvigorate the sense of shared responsibility and social justice among stakeholders within a healthcare delivery system. This renewed commitment can, in turn, help restore trust, truthfulness, and legitimacy between the healthcare delivery system and the population it serves.

A critical element to consider is that stewardship thrives in the context of a multicultural society, where diverse perspectives contribute to shaping equitable and inclusive healthy public policies. At the same time, it often requires making difficult decisions that may not initially be popular but serve the underlying best interests of the population. These decisions lie at the heart of healthcare stewardship, reflecting its core principles. Importantly, the path to improved healthcare outcomes under the stewardship framework is seldom smooth or linear; it demands constant vigilance to identify and address competing self-interests among various stakeholders.

Building partnerships is one of healthcare stewardship's central functions, and its potential lies in its ability to revive the concept of the commons—a shared set of values and collective responsibility. This modern reinterpretation of the commons (Cassel et al., 2007; Hiatt, 1975; Hardin, 1968) emphasizes the importance of collaboration and ethical governance in healthcare services delivery. At its core, healthcare stewardship is grounded in a multilevel socio-ecological contract between the different levels of healthcare delivery system organization and the population it serves.

By upholding this multilevel socio-ecological contract and embracing the principles of the commons, healthcare stewardship transcends the narrow confines of cost efficiency and market-driven health economics. Instead, it establishes a socially responsible and normatively guided framework that focuses on achieving positive health outcomes that are equitable, ethical, and aligned with the broader values of society. Through such stewardship, healthcare delivery systems around-the-world can address pressing challenges while fostering trust and equity within their communities.

A stewardship approach, based in developing the collective health commons, fits well with the mission of providing sound empirically-driven evidence-based clinical services by health care providers. For physicians, this mission-oriented framework lies at the core of the Hippocratic Oath. The concept of health care stewardship is also consistent with value-based medicine and empirically-supported clinical practice guidelines. Grounded in both ethical and economic norms, the healthcare services delivery strategy based on stewardship can provide the available evidence about what works well to support population-based measures that can improve overall health, wellness, well-being, and resilience.

Applying the principles of healthcare stewardship to present-day issues facing the global healthcare delivery system creates the careful, well-constructed opportunity of managing limited health care resources in a transparent, accountable, and socially-responsible manner. Stewardship applies economically-driven interests in the evidence-based delivery of health care goods and services by balancing those interests with an ethically normative collection of values (i.e., the medical commons). This allows for the overall health, wellness, well-being, and resilience of the population being served by the healthcare delivery system to take priority, while the healthcare system fulfills its operational, political, and social mission and vision. There is much more work to be done in the quantitative assessment of stewardship function. Additional research and identification of other domains that fully delineate stewardship activity is indicated.

2.6 References

1. Abrams, E. M., Singer, A. G., Shaker, M., et al. (2021). What the COVID-19 pandemic can teach us about resource stewardship and quality in health care. *The Journal of Allergy and Clinical Immunology: In Practice 9*(2): 608-612.
2. Brinkerhoff, D.W. and Bossert, T.J. (2014) Health Governance: Principal-Agent Linkages and Health System Strengthening. *Health Policy and Planning*, 29, 685-693.
3. Brinkerhoff, D.W., Cross, H.E., Sharma S., et al. (2019a) Stewardship and health systems strengthening: An overview. *Public Admin Dev* 39: 4–10.
4. Brinkerhoff, D. W., & Bossert, T. J. (2019b). *Health System Stewardship: Governance and institutional dynamics.* World Health Organization.
5. Cassel, C. K, and T. E. Brennan. 2007. Managing medical resources: Return to the commons? *JAMA* 297 (22): 2518–2521.
6. Dion, H., Evans, M. (2024). Strategic frameworks for sustainability and corporate governance in healthcare facilities; approaches to energy-efficient hospital management. *Benchmarking: An International Journal* 31(2): 353-390.
7. Fatemi, Y., Bergl, P. A. (2022). Diagnostic stewardship: Appropriate testing and judicious treatments. *Critical Care Clinics* 38(1): 69-87.
8. Greer, S.L. (2018). Organization and Governance: Stewardship and Governance in Health Systems. In: van Ginneken, E., Busse, R. (eds) Health Care Systems and Policies. Health Services Research. Springer, New York, NY.
9. Guidolin, K., Catton, J., Rubin, B., et al. (2022) Ethical decision making during a healthcare crisis: a resource allocation framework and tool. *Journal of Medical Ethics* 48(8): 504-509.
10. Hardin, G. (1968) Tragedy of the Commons. Science 162(3859): 1243-1248.

11. He, A. J., Bali, A. S., Ramesh, M. (2022). Active stewardship in healthcare: Lessons from China's health policy reforms. *Social Policy & Administration* 56(6), 925-940.
12. Hiatt, H.H. (1975) Protecting the Medical Commons: Who Is Responsible? New England Journal of Medicine, 293, 235-241.
13. Hu, H., Cohen, G., Sharma, B., et al. (2022). Sustainability in health care. *Annual Review of Environment and Resources* 47(1): 173-196.
14. Hubbard, R., Paquet, G., Wilson, C. (2022). *Stewardship: Collaborative Decentred Metagovernance and Inquiring Systems.* Ottawa, Canada: University of Ottawa Press.
15. Institute of Medicine (IOM) Committee on Quality of Health Care in America. (2001). *Crossing the Quality Chasm: A New Health System for the 21st Century.* National Academies Press, Washington, DC.
16. Kaiser, R. A., Taing, L., Bhatia, H. (2022) Antimicrobial resistance and environmental health: a water stewardship framework for global and national action *Antibiotics* 11(1): 63.
17. Kapoor, N., Kumar, D., Thakur, N. (2014). Core attributes of stewardship; foundation of sound health system. *International journal of health policy and management 3*(1): 5–6.
18. Khatri, R.B., Wolka, E., Nigatu, F. *et al.* (2023) People-centred primary health care: a scoping review. *BMC Prim. Care*, 24, 236.
19. MacNeill, A. J., McGain, F., Sherman, J. D. (2021). Planetary health care: a framework for sustainable health systems. *The Lancet Planetary Health 5*(2): e66-e68.
20. Meagher, K. M., Watson, S., Suh, G. A., & Virk, A. (2022). The new precision stewards? *Journal of Personalized Medicine* 12(8): 1308.
21. Redvers, N. (2021). Patient-planetary health co-benefit prescribing: emerging considerations for health policy and health professional practice. *Frontiers in Public Health 9*: 678545.
22. Robu, D., Lazar, J. B. (2020). Shaping Successful Stakeholder Engagement by Design: Digital Transformation in Healthcare.

In European Conference on Knowledge Management. Academic Conferences International Limited.

23. Saltman, R. B., and O. Ferroussier-Davis. (2000) The concept of stewardship in health policy. *Bulletin of the World Health Organization* 78(6): 732–739.
24. Scott, R. E., Morris, C., Mars, M. (2021) Development of a "cellphone stewardship framework": legal, regulatory, and ethical issues. *Telemedicine and e-Health* 27(3): 316-322.
25. Shaw, E., Walpole, S., McLean, M., et al. (2021). AMEE Consensus Statement: Planetary health and education for sustainable healthcare. *Medical Teacher* 43(3): 272-286.
26. Stanford, J. (2023). *Promoting health equity through evidence-based healthcare stewardship.*
27. Tello, J. E., Barbazza, E., Waddell, K. (2020). Review of 128 quality of care mechanisms: A framework and mapping for health system stewards. *Health Policy* 124(1): 12-24.
28. Travis, P, Egger, D, Davies, P et al. (2002). Towards better stewardship: concepts and critical issues. World Health Organization.
29. White-Williams, C., Bittner, V., Eagleson, R., et al. (2022). Interprofessional collaborative practice improves access to care and healthcare quality to advance health equity. *The Journal for Healthcare Quality (JHQ)* 44(5), 294-304.
30. Wilson, T., Bevan, G., Gray, M., et al. (2020) Developing a culture of stewardship: how to prevent the Tragedy of the Commons in universal health systems. *Journal of the Royal Society of Medicine* 113 (7): 255-261.
31. Woolhandler, S., Himmelstein, D. U., Ahmed, S., et al. (2021) Public policy and health in the Trump era. *The Lancet* 397(10275): 705-753.
32. World Health Organization (WHO). (2020). *Improving governance and resource management in health systems.*

3.0

Foundations, Capacity Building, and Sustainable Strategies for Healthcare Stewardship.

3.1 Challenges Facing Healthcare Stewardship.

Healthcare stewardship faces numerous challenges that can impede the effective management, allocation, and utilization of limited healthcare resources.

1. *Resource Limitations.*
   a) *Funding Constraints*: Limited financial resources significantly hinder the availability of essential healthcare services and medications, particularly in low-income countries and underserved areas. These financial constraints often lead to reduced access to preventive care, diagnostic services, and life-saving treatments, exacerbating health inequities. Budget cuts and economic downturns further compound these challenges, forcing healthcare delivery systems in low-income countries to operate with minimal healthcare resources, which can compromise the quality and sustain-

ability of care. According to a scoping review published in the International Journal for Equity in Health, economic barriers disproportionately affect vulnerable populations, leading to catastrophic health expenditures and widening disparities in healthcare access and outcomes (Teixeira de Siqueira Filha et al., 2022).

b) *Workforce Shortages*: The shortage of healthcare professionals—including doctors, nurses, and allied health workers—poses significant challenges to healthcare delivery systems worldwide. These shortages often result in heavier workloads for existing staff, leading to physical and emotional exhaustion, commonly known as burnout. This, in turn, can compromise the quality of care delivered to patients and strain the resilience of healthcare systems. A 2021 report by the World Health Organization (WHO) highlights that the global shortage of healthcare workers, particularly in low- and middle-income countries, exacerbates disparities in healthcare access and outcomes, making it a critical issue to address for sustainable healthcare delivery.

2. *Inequitable Access*

a) *Disparities in Healthcare*: Upstream determinants of health—including income, education, housing stability, food security, transportation, and geographic location—play a pivotal role in shaping access to healthcare services delivery. These factors often create significant disparities in navigation and utilization, disproportionately affecting marginalized and vulnerable populations. Individuals in these groups frequently encounter extreme obstacles, such as limited financial resources, geographic isolation, and systemic discrimination, which hinder their ability to receive high-quality and safe medical care. Addressing these upstream determinants is essential for reducing health inequities and fostering equitable healthcare systems. According to a study published in the *International Journal for Equity in Health*, the intersection of health-related social needs (HRSNs) and structural barriers perpetuates disparities in healthcare access, emphasizing the need for targeted interventions to improve health outcomes (Teixeira de Siqueira Filha et al., 2022).

b) *Rural and Remote Areas*: Delivering healthcare services in rural and remote areas presents a complex set of challenges. Logistical issues, such as long travel distances and limited transportation options, often hinder access to care. Additionally, the lack of infrastructure—including inadequate medical facilities and unreliable communication networks—further compounds these difficulties. These regions frequently face a shortage of well-trained healthcare professionals, which limits the availability of specialized care and places immense pressure on existing staff. Addressing these challenges requires targeted investments in infrastructure, workforce development, and innovative solutions like telemedicine to bridge the gap in healthcare access. According to the U.S. Government Accountability Office (GAO), rural areas experience higher rates of hospital closures and provider shortages, emphasizing the need for systemic interventions to improve healthcare delivery in these communities (GAO, 2023).

3. *Policy and Governance Issues.*

a) *Inconsistent Policies*: Variability in healthcare-related public policies across regions or countries introduces significant challenges to the delivery of care. Confusion stemming from inconsistent regulations can lead to inefficiencies in healthcare delivery systems, as providers struggle to navigate differing requirements and standards. This variability often hampers the implementation of evidence-based best clinical practices, as healthcare professionals may face conflicting guidelines or limited resources to adapt to diverse policy frameworks. Addressing these inconsistencies is crucial for fostering a cohesive approach to healthcare services delivery and ensuring equitable access to high-quality care. A study published in the *International Journal for Quality in Health Care* highlights the impact of unwarranted variation in clinical practices, emphasizing the need for harmonized policies to reduce disparities and improve patient outcomes (Atsma et al., 2020).

b) *Corruption and Mismanagement*: Corruption and mismanagement in healthcare funding and resource allocation pose significant threats to the integrity and effectiveness of healthcare delivery systems. Fraud, waste, and abuse of funds and resources not only divert critical resources away from patient care but also undermine the efficiency and sustainability of healthcare delivery. These practices often result in reduced quality and safety of care, as healthcare providers struggle to operate within compromised systems. Furthermore, corruption erodes public trust in healthcare institutions, discouraging individuals from seeking necessary medical services and exacerbating health inequities. According to the National Academies of Sciences, Engineering, and Medicine, approximately $455 billion of global healthcare expenditures are lost annually to fraud and corruption, highlighting the urgent need for transparent governance and robust accountability measures to safeguard healthcare systems (National Academies of Sciences, Engineering, and Medicine, 2018).

4. *Technological Barriers.*

   a) *Digital Divide*: The adoption of health information technology (HIT), including electronic health records (EHRs), SMART devices, broadband access, and telemedicine, holds immense potential to transform healthcare services delivery. However, its implementation is often hindered by significant barriers, particularly in low-resource settings. A lack of robust digital infrastructure, such as reliable internet connectivity and modern hardware, poses a critical challenge. Additionally, the shortage of technical expertise among healthcare providers and administrators further complicates the integration of these technologies into healthcare structures and processes. These limitations not only delay the adoption of HIT but also exacerbate existing disparities in healthcare access and quality. Addressing these challenges requires targeted investments in infrastructure development, workforce training, and capacity-building initiatives. According to a review published in *BMC Medical Informatics and Decision Making*, the successful implementation of HIT in resource-constrained environments depends on a comprehensive approach that includes stakeholder engagement, tailored solutions, and ongoing support to ensure sustainability (Jawhari et al., 2016).

b) *Personal Security Concerns*: As healthcare delivery systems around-the-globe increasingly rely on digital tools, protecting patient privacy and safeguarding personal identifiable information have become critical challenges. The integration of electronic health records (EHRs), telemedicine platforms, and Internet of Medical Things (IoMT) devices has expanded the digital footprint of healthcare, exposing information systems to a growing array of cyber threats. These threats include ransomware attacks, phishing schemes, and data breaches, which can compromise sensitive patient information and disrupt essential healthcare services. Ensuring robust cybersecurity measures, such as encryption, multi-factor authentication, and regular system audits, is essential to mitigate these risks. However, many healthcare organizations face resource constraints and a lack of specialized expertise, particularly in low-resource settings, which further complicates efforts to secure their systems.

According to a report by the U.S. Department of Health and Human Services (HHS), healthcare remains one of the most targeted industries for cyberattacks, underscoring the urgent need for comprehensive strategies to protect patient data and maintain trust in global healthcare delivery systems (HHS, 2023).

c) *Interoperability*: Enhancing the interoperability of health information technology (HIT) and health information systems (HIS) is essential for fostering seamless, bi-directional communication between independent healthcare entities. Effective interoperability enables healthcare providers to access and exchange patient data in a coordinated and meaningful manner, improving care continuity and reducing inefficiencies. However, achieving this requires overcoming significant barriers, such as disparate data standards, legacy systems, and organizational resistance to change. Standardized frameworks like HL7 FHIR (Fast Healthcare Interoperability Resources) and robust data-sharing protocols are critical for addressing these challenges. Additionally, investments in infrastructure, workforce training, and stakeholder collaboration are necessary to ensure the successful implementation of interoperable systems. According to a report by the Office of the National Coordinator for Health Information Technology (ONC), approximately 44% of hospitals in the United States face challenges in exchanging patient information, underscoring the urgent need for coordinated efforts to improve HIT interoperability (ONC, 2021).

5. *Resistance to Change.*

a) *Cultural and Organizational Resistance*: Healthcare organizations and providers often exhibit resistance to changes in established medical practices and clinical protocols, which can significantly hinder the adoption of new, evidence-based, and empirically-supported approaches to medical care. This resistance may stem from various factors, including a reluctance to deviate from familiar routines, concerns about the costs and resources required for implementation, and skepticism regarding the efficacy of new practices. Additionally, organizational inertia and a lack of leadership support can further impede progress, creating barriers to innovation and continuous quality improvement. Overcoming these challenges requires targeted strategies, such as fostering a culture of continuous learning, providing education and training on empirically-driven evidence-based clinical practices, and engaging stakeholders in the decision-making process. According to a study published in the *International Journal of Health & Medical Sciences*, addressing these barriers through organizational support and access to evidence-based resources is crucial for promoting widespread adoption and enhancing patient care outcomes (Aishammari et al., 2020).

b) *Patient Non-Adherence*: Ensuring patient adherence to treatment plans for chronic conditions, as well as promoting disease prevention through early detection and screening measures, presents significant challenges. Factors such as limited health literacy, socioeconomic barriers, and a lack of trust in healthcare systems can hinder patients from following prescribed regimens or participating in preventive care initiatives. Non-adherence not only compromises individual health outcomes but also places a strain on limited healthcare resources, leading to increased hospitalizations, higher treatment costs, and preventable disease progression. Addressing these challenges requires a multilayered approach, including patient education, personalized care plans, and the integration of digital health tools to support adherence. According to a review published in *Medicina*, improving adherence through targeted interventions can enhance clinical outcomes, reduce healthcare costs, and improve overall patient quality of life (Religioni et al., 2025).

6. *Public Health Emergencies.*
   a) *Pandemics and Epidemics*: Public health emergencies—including epidemics, pandemics, and endemics—place immense strain on healthcare delivery systems, disrupting both traditional medical care and essential public health services. These crises often lead to resource shortages, overwhelmed healthcare facilities, and increased morbidity and mortality rates. The surge in demand for medical care during such emergencies can exhaust healthcare workers, deplete medical supplies, and compromise the quality of care provided to patients. Additionally, the ripple effects of these emergencies extend beyond direct healthcare delivery, impacting public health functions such as disease surveillance, vaccination programs, and health education initiatives. Addressing these challenges requires robust preparedness measures, including investments in healthcare infrastructure, workforce resilience, and coordinated response strategies. According to a study published in *Archives of Public Health*, countries with weak health systems and inadequate preparedness face heightened vulnerabilities during public health emergencies, emphasizing the need for integrated approaches to mitigate their impact and enhance health security (Khatri et al., 2023).
   b) Natural Disasters: Natural disasters, such as hurricanes, earthquakes, floods, and wildfires, can severely disrupt healthcare services delivery and damage critical infrastructure, including hospitals, clinics, and supply chains. These events often create urgent healthcare needs, such as treating injuries, managing infectious disease outbreaks, and addressing mental health crises, which place immense strain on existing resources and capabilities. The destruc-

tion of infrastructure and the displacement of populations further exacerbate these challenges, making it difficult to deliver timely and effective care. Additionally, the long-term impacts of natural disasters can include reduced access to healthcare services, worsened health disparities, and increased vulnerability to future emergencies. According to the National Academies of Sciences, Engineering, and Medicine, enhancing the resilience of healthcare delivery systems through investments in infrastructure, workforce training, and disaster preparedness is essential to mitigate the effects of natural disasters and ensure continuity of care during crises (National Academies of Sciences, Engineering, and Medicine, 2025).

7. *Moral and Ethical Dilemmas.*
   a) *Resource Allocation Decisions*: Decisions regarding the allocation of limited healthcare resources, such as organ transplantation or the distribution of scarce high-cost medications (e.g., cell gene therapy), present profound ethical dilemmas. These decisions require a delicate balance between competing principles, including justice, equity, and utility. The challenge lies in ensuring that resources are distributed fairly while maximizing the overall benefit to society. For instance, prioritizing patients based on medical urgency, likelihood of success, or societal contribution can lead to difficult trade-offs and potential perceptions of bias. Furthermore, disparities in access to healthcare, influenced by socioeconomic status, geographic location, and systemic inequities, complicate efforts to achieve equitable resource distribution. Addressing these challenges necessitates transparent decision-making frameworks that incorporate ethical principles and engage diverse stakeholders. According to a study published in *BMC Medical Ethics*, ethical resource allocation during public health crises requires a balance of

equity, reciprocity, and proportionality to ensure that decisions are both just and effective (O'Sullivan et al., 2022).

b) *Balancing Cost and Quality*: Balancing cost-saving measures with the imperative to maintain the quality and safety of healthcare services remains a persistent and complex challenge in healthcare stewardship. Efforts to reduce expenses, such as streamlining operations, optimizing resource allocation, or implementing new technologies, often carry the risk of unintended consequences, including diminished patient outcomes or compromised care standards. For instance, staff reductions or limitations on essential supplies may lead to increased workloads for healthcare providers, potentially affecting their ability to deliver safe and effective care. Addressing this challenge requires a strategic approach that integrates cost-efficiency with empirically-driven evidence-based clinical practices and robust quality assurance mechanisms. According to a study published in the *Journal of Healthcare Management*, successful cost-containment strategies must prioritize patient-centered care, involve multidisciplinary collaboration, and incorporate continuous monitoring to ensure that financial sustainability does not come at the expense of healthcare quality (Kaplan & Haas, 2025).

8. *Global Health Issues.*

a) *Cross-Border Health Threats*: Emerging infectious diseases (EIDs) and other communicable health threats that transcend national borders pose significant challenges to global health security and demand coordinated international responses from healthcare delivery systems worldwide. The complexity of achieving such coordination lies in the disparities among nations in terms of priorities, resources, and operational capabilities at multiple levels of governance. Addressing these challenges requires an integrated approach that aligns international health objectives, improves cross-border communication, and strengthens healthcare system infrastructures to respond effectively to these threats. Global healthcare delivery systems often face obstacles due to varying levels of development, cultural differences, and political interests, which can impede effective collaboration. For example, countries with limited health infrastructure may struggle to implement advanced containment measures or early warning systems, highlighting inequities that exacerbate the difficulty of managing cross-border health threats (Fidler, 2017). Furthermore, differing priorities—such as economic growth versus health investments—can create tensions and hinder the establishment of common strategies. Multilateral organizations such as the World Health Organization (WHO) and regional health alliances play crucial roles in fostering coordination and providing technical support to countries facing infectious disease outbreaks (World Health Organization, 2020). Enhanced cooperation in research, surveillance systems, and resource sharing is essential to bridge gaps in capabilities and ensure a collective global response to emerging health threats. Strengthening the global healthcare delivery system responses requires not only technical advancements but also political commitment to equity and resilience. By focusing on capacity building, fostering trust among nations, and addressing systemic inequities, the global community can move toward a more unified and effective approach to combating transnational health threats.

b) *Health Inequities*: Addressing global health inequities and ensuring equitable access to essential healthcare services for all populations remains a multifaceted and enduring challenge. Persistent disparities in healthcare availability, quality, and affordability are exacerbated by factors such as economic inequality, geographic barriers, cultural differences, and systemic inequities within and across nations.

These issues are particularly pronounced in low- and middle-income countries, where healthcare infrastructure often lacks the capacity to meet population needs, and where marginalized groups face significant barriers to accessing care. Efforts to tackle global health inequities must involve collaborative international initiatives aimed at improving multilevel healthcare delivery systems, fostering innovation, and prioritizing equity as a foundational principle. For instance, organizations such as the World Health Organization (WHO) and the Global Fund to Fight AIDS, Tuberculosis, and Malaria have been instrumental in developing frameworks for equitable healthcare access and providing resources to underserved regions (World Health Organization, 2022). In addition, partnerships between governments, non-governmental organizations (NGOs), and private sectors can play a crucial role in enhancing health equity through investments in infrastructure, education, and preventive care. Addressing systemic challenges requires a commitment to long-term solutions, such as increasing healthcare funding, expanding universal healthcare coverage, and integrating community-based approaches to ensure that vulnerable populations are not left behind. Moreover, advancing equity in healthcare requires the acknowledgment and dismantling of structural inequalities, including those rooted in social determinants of health. By leveraging data-driven approaches, policy advocacy, and grassroots community engagement, the global health community can strive to mitigate disparities and promote a more sustainable and equitable healthcare system for future generations.

These challenges highlight the need for continuous efforts to improve healthcare stewardship through creative and innovative solutions, collaboration and cooperation among global stakeholders, and the development of robust healthy public policies and practices that promote the sustainable and equitable use of healthcare resources.

3.2 Foundational Pillars of Healthcare Stewardship.

Healthcare stewardship involves the responsible planning, allocation, and management of limited health resources (Lear et al., 2016). The foundational characteristics of healthcare stewardship are many.

Transparency and accountability are essential pillars for fostering trust among those served by healthcare delivery systems. Trust is built when healthcare providers, administrators, and policymakers operate with openness and demonstrate responsibility for their actions, decisions, and the impact of those choices on patient care and overall healthcare system performance. This requires that all stakeholders—at every level of the healthcare delivery system—are held answerable for their contributions to care delivery and resource stewardship. Establishing transparent structures and processes for resource allocation and decision-making is imperative for achieving equitable and ethical healthcare delivery. Such processes must be built on principles of cultural humility and empathy, recognizing the diverse needs and values of patients and communities. Cultural humility involves an ongoing commitment to self-reflection and to understanding the cultural contexts and experiences of others, fostering equitable and compassionate care. Empathy strengthens this foundation by ensuring that decision-making prioritizes the well-being of individuals and communities over bureaucratic convenience or financial expediency (Turk, 2020). Clear, truthful, and consistent communication is a cornerstone of transparency. Providing all stakeholders—patients, providers, and policymakers—with access to comprehensive and comprehensible information about healthcare policies, procedures, and outcomes is vital for ensuring informed decision-making and participatory governance. For instance, making data on how resources are allocated and utilized accessible to the public builds credibility and reinforces the integrity of healthcare systems. Similarly, encouraging stakeholder involvement in decision-making processes fosters mutual trust and shared accountability. Accountability further strengthens these efforts by ensuring that mechanisms are in place to evaluate performance, correct errors, and acknowledge successes. Robust accountability frameworks should include key performance metrics, ethical standards, and operational systems for addressing grievances, thereby creating an environment of continuous quality improvement (Institute of Medicine, 2001). Together, transparency and accountability help establish healthcare delivery systems that are trustworthy, ethical, and capable of meeting the needs of the populations they serve. By embedding these principles into healthcare services delivery, healthcare organizations and systems can build lasting trust, improve the quality and safety of care, and ensure the sustainability of global healthcare delivery systems in the long term.

Moral and ethical decision-making is foundational to ensuring that healthcare decisions regarding the allocation of limited resources are guided by ethical principles such as justice, autonomy, beneficence, and non-maleficence. These principles provide a framework for equitable and fair decision-making processes, prioritizing the distribution of healthcare goods and services in a manner that upholds patient rights, dignity, and well-being. The principle of justice emphasizes the fair allocation of resources, ensuring that all individuals, regardless of socioeconomic status, ethnicity, or geographic location, have access to essential healthcare services. Autonomy highlights the importance of respecting patients' rights to make informed decisions about their own care, fostering trust and empowerment in the patient-provider relationship (Beauchamp & Childress, 2019). Beneficence focuses on promoting the best possible outcomes for patients by actively providing benefits and care that improve their health and welfare. Non-maleficence, on the other hand, ensures that harm is avoided, advocating for the prevention of practices or policies that could negatively impact patients or exacerbate disparities. In addition to these ethical foundations, prioritizing patient health, safety, and welfare (HSW) is critical in fostering high-quality healthcare service delivery. Emphasizing health equity seeks to address systemic disparities by improving access to care for marginalized populations, reducing barriers such as financial constraints or geographic limitations. Enhancing health literacy—by equipping individuals with the knowledge and skills to navigate healthcare systems effectively—further contributes to equitable outcomes, empowering patients to make informed decisions and advocate for their own health needs (World Health Organization, 2020).

Operational efficiency and cost-effectiveness in healthcare service delivery are essential to optimizing resource utilization, improving health outcomes, and mitigating issues such as fraud, waste, and abuse. Achieving these goals requires a strategic focus on streamlining processes, integrating technology, and fostering transparency within healthcare delivery systems to ensure that resources are allocated effectively and equitably. Efficiency in healthcare delivery encompasses measures such as reducing redundancies, improving coordination of care, and implementing empirically-driven evidence-based clinical practices to minimize unnecessary interventions. By enhancing operational workflows, healthcare organizations can provide higher-quality care while ensuring that expenditures are aligned with the desired outcomes (Institute of Medicine, 2013). For example, adopting digital health solutions, such as electronic health records (EHRs) and telemedicine platforms, can improve care coordination, reduce administrative burdens, and facilitate better tracking of patient outcomes. Cost-effectiveness, on the other hand, emphasizes the importance of achieving maximum health benefits relative to the costs incurred. This involves analyzing the value of interventions, treatments, and diagnostic tools to prioritize those that yield the greatest impact on population health within constrained budgets (World Health Organization, 2021). By employing cost-effectiveness analyses, policymakers and healthcare providers can make informed decisions that balance financial sustainability with improved care delivery. Moreover, addressing fraud, waste, and abuse within healthcare delivery systems is critical to maintaining financial integrity and trust. Proactive measures such as robust auditing, monitoring systems, and compliance programs can help detect and prevent fraudulent practices, as well as reduce inefficiencies caused by wasteful spending or improper billing. Through a commitment to operational efficiency and cost-effectiveness, global healthcare systems can not only optimize resource use but also promote equity, improve patient outcomes, and ensure long-term sustainability.

Integrating environmental sustainability into healthcare service delivery and resource use is essential to ensuring that current and future generations of patients benefit from effective, equitable, and resilient healthcare delivery systems. Long-term environmental sustainability must be a cornerstone of any strategic plan, incorporating approaches that balance resource allocation and utilization with the preservation of essential assets such as environmental quality, financial stability, and community health. By designing healthcare delivery systems that prioritize environmentally sustainable operations, organizations can simultaneously improve outcomes and reduce their ecological footprint. In this context, addressing the environmental determinants of health includes recognizing the impacts of healthcare practices on human-induced climate change, waste management, and resource consumption. For example, hospitals can reduce their carbon footprint by implementing energy-efficient infrastructure, sustainable medical supply chains, and low-waste practices. These actions not only preserve environmental resources but also contribute to healthier communities by mitigating pollution and other environmental hazards (Pichler et al., 2019). Economic sustainability focuses on maximizing cost-efficiency while ensuring adequate funding for essential health services. This requires evaluating the long-term financial implications of healthcare investments, optimizing resource allocation, and preventing economic disparities in healthcare access. Commercial determinants, such as the influence of private sector interests and pharmaceutical pricing, also play a significant role and must be managed to align with public health priorities rather than profit motives (Kickbusch et al., 2016). Health-related social needs (HRSNs), including education, stable housing, food security, transportation, and employment, are equally critical in shaping health outcomes. Addressing these determinants involves upstream interventions that improve living conditions, enhance health literacy, and reduce inequalities that contribute to poor health. A comprehensive, 360-degree approach to health determinants (e.g., community health assessments) allows global healthcare delivery systems to understand and address the interconnected needs of individuals and communities, fostering more holistic, integrated, comprehensive, and impactful healthcare strategies. By embedding the principles and practices of environmental sustainability into the strategic vision of healthcare delivery systems and addressing the broader determinants of health, we can move toward a model that supports not only the well-being of individuals but also the resilience and equity of society at large.

Quality improvement science should serve as a pivotal framework in efforts to enhance the quality and safety of healthcare service delivery. By employing empirically-derived tools and evidence-based clinical practice guidelines, this scientific discipline provides structured methodologies that ensure healthcare delivery systems are continuously evolving to meet patient needs effectively. Furthermore, encouraging disruptive innovation within healthcare sectors that are most in need can lead to transformational changes, fostering healthcare systems that are more efficient, equitable, and capable of delivering superior health outcomes. Central to quality improvement science is the application of data-driven approaches to evaluate and refine healthcare processes. Empirically-derived tools such as Plan-Do-Study-Act (PDSA) cycles, Six Sigma methodologies, and Lean frameworks offer systematic methods for identifying inefficiencies, assessing interventions, and sustaining improvements over time (Institute for Healthcare Improvement, 2021). These tools empower healthcare providers to implement measurable changes that are aligned with both patient safety and organizational goals. The implementation of evidence-based clinical practice guidelines is another cornerstone of quality improvement science. By translating clinical research and proven methodologies into actionable clinical strategies, global healthcare delivery systems can reduce variability in healthcare services delivery and enhance patient outcomes. For example, empirically driven evidence-based protocols for infection control have dramatically lowered rates of healthcare-associated infections in many settings around the globe (World Health Organization, 2020). Disruptive innovation, as an essential complement to traditional quality improvement science, focuses on addressing critical gaps in healthcare services delivery. Innovations such as telemedicine, portable diagnostic devices, and artificial intelligence in clinical decision-making have revolutionized access to care, particularly in underserved or remote populations. Encouraging such innovations not only improves service delivery efficiency but also challenges the status quo, fostering creative solutions to persistent healthcare challenges (Christensen et al., 2009). By integrating empirically-derived tools, evidence-based clinical practice guidelines, and disruptive innovation, quality improvement science provides a comprehensive framework for transforming healthcare service delivery and healthcare delivery systems. This approach ensures that systems remain adaptable to emerging challenges while prioritizing patient safety, efficiency, and equitable access to care.

Guaranteeing equitable access to healthcare services delivery that individuals need, prefer, and value—irrespective of socio-economic status, ethnicity, or geographic location—is pivotal to addressing health disparities, improving health literacy, and fostering inclusive, empirically-driven evidence-based healthy public policies. Achieving this goal requires intentional and sustained efforts across healthcare systems, governments, and communities to dismantle barriers that prevent individuals from receiving care tailored to their unique circumstances and preferences. Health disparities persist as a significant challenge globally, disproportionately impacting marginalized populations who often face obstacles such as economic hardship, lack of transportation, limited health infrastructure, and systemic bias (World Health Organization, 2020). To overcome these disparities, healthcare delivery systems must prioritize equitable resource allocation, ensuring that underserved regions and vulnerable groups receive adequate healthcare services. For example, expanding healthcare infrastructure in rural or low-income areas, coupled with subsidized healthcare delivery programs, can bridge gaps in access and reduce inequities. Improving health literacy is equally critical in empowering individuals to make informed decisions about their care. Health literacy initiatives should focus on providing clear, culturally sensitive, and accessible information that resonates with diverse populations, enabling them to navigate healthcare systems efficiently and effectively (Nutbeam, 2008). Programs that promote health education in schools, community centers, and workplaces can significantly enhance awareness and understanding of healthcare services options, encouraging and focusing on proactive more than reactive healthcare services. Inclusive healthy public policies serve as the backbone of efforts to achieve equitable healthcare access. Policymakers must collaborate with diverse stakeholders, including community leaders, healthcare providers, and patients themselves, to develop strategies that reflect the real-world needs, preferences, and values of populations. Such policies should be grounded in social justice, equity, and transparency, ensuring that they address all of the determinants of health that shape individual and community well-being (Braveman et al., 2011). By integrating approaches to reduce disparities, enhance health literacy, and build inclusive healthy public policies, healthcare delivery systems can establish equitable access to care as a foundational principle. These efforts not only improve individual and population health outcomes but also foster societal trust and developed resilience in the face of evolving healthcare challenges.

The 4 C's—collaboration, cooperation, coordination, and communication—along with capacity building are essential and indispensable for fostering effective partnerships among diverse stakeholders, including government agencies, healthcare providers, patients, and communities. These elements provide the foundation for addressing the broader determinants of health, leveraging cross-disciplinary discourse and actionable solutions grounded in a socio-ecological framework across macro, meso, and micro levels of organization. **Collaboration** entails a synergistic approach where stakeholders actively work together to pool resources, expertise, and perspectives to achieve common health objectives. This collaborative effort is pivotal in tackling complex health challenges, such as health inequities or systemic barriers, by integrating diverse disciplines and fostering innovation. For example, partnerships between healthcare organizations and local governments can result in targeted programs addressing health-related social needs (HRSNs) like stable and safe housing or food security (Rosen et al., 2018). **Cooperation** emphasizes the importance of mutual support and shared responsibility among stakeholders. By nurturing cooperative relationships, healthcare delivery systems can break down silos, align goals, and harmonize efforts, fostering a cohesive and unified response to public health issues. Cooperation is especially critical during crises, such as disease outbreaks, where collective action ensures timely interventions and resource sharing (World Health Organization, 2021). **Coordination** focuses on streamlining processes, aligning strategies, and reducing redundancies to ensure efficiency in healthcare delivery. Effective coordination ensures that efforts across different levels of organization—macro (national policies), meso (regional health systems), and micro (individual care)—are aligned to support sustainable health outcomes. For instance, coordination among healthcare providers ensures seamless transitions of care, avoiding gaps in patient treatment and improving outcomes (Institute of Medicine, 2003). **Communication** serves as the backbone of the 4 C's, enabling stakeholders to exchange critical information, share best practices, and foster understanding across diverse perspectives. Transparent and effective communication not only builds trust but also empowers patients and communities to actively participate in shaping their healthcare experiences. Clear communication also aids in conveying healthy public policies and educating the public about proactive care strategies (Nutbeam, 2008). Capacity building amplifies the impact of the 4 C's by equipping stakeholders with the skills, infrastructure, and resources needed to address complex health challenges. Investing in capacity building ensures that healthcare systems remain resilient and adaptable, capable of responding to evolving needs while maintaining equity and sustainability (Braveman et al., 2011). Through the integration of the 4 C's and capacity building within a socio-ecological framework, healthcare delivery systems world-wide can address the upstream determinants of health at all levels of organization. At the macro level, this involves national policy interventions aimed at reducing systemic inequalities. At the meso level, regional efforts may focus on improving health infrastructure and community engagement. At the micro level, individual care is personalized and culturally sensitive, addressing unique needs, preferences, and values. This holistic and integrated approach provides a comprehensive pathway to sustainable, equitable healthcare outcomes and strengthens the collective ability of stakeholders to address the broader determinants of health effectively.

Establishing robust regulatory and statutory frameworks is vital to overseeing healthcare services delivery and healthcare delivery systems whilst ensuring compliance with the highest global quality standards. These frameworks serve as the foundation for promoting ethical conduct, improving patient safety, and delivering equitable healthcare services to individuals and populations served. Effective regulations provide a structured and enforceable system to monitor performance, minimize risks, and foster accountability among healthcare organizations and professionals. Regulatory frameworks must encompass key components such as accreditation systems, key performance improvement benchmarks, and mechanisms for addressing non-compliance. Accreditation ensures that healthcare facilities meet established standards and operate in alignment with empirically-driven evidence-based clinical practice guidelines, while benchmarks help measure outcomes and promote continuous quality improvement (Institute of Medicine, 2003). Additionally, enforcing regulations can prevent malpractice, enhance patient protections, and maintain public confidence in healthcare institutions. Implementing effective leadership and governance structures is equally crucial to managing healthcare delivery systems efficiently and ethically. Leadership teams must prioritize transparency, accountability, and truthfulness in their decision-making processes, creating a culture of integrity that permeates all levels of the organization. Governance structures should be designed to oversee resource allocation, uphold ethical standards, and evaluate system performance while engaging stakeholders across the service community. Transparency within healthcare delivery systems involves open communication about policies, procedures, and outcomes. Providing clear, truthful information to patients, providers, and other stakeholders builds trust and ensures informed decision-making. For example, public reporting of healthcare outcomes and financial expenditures reinforces accountability and increases confidence in leadership (Turk, 2020). Accountability frameworks, such as key performance metrics and reporting systems, enable healthcare organizations to identify areas for improvement and demonstrate responsibility for actions taken. Leaders who prioritize accountability foster environments that encourage continuous learning and adaptation, ensuring sustained excellence in care delivery. By establishing transparent, accountable governance processes and adhering to robust regulatory frameworks, healthcare delivery systems can build trust within their service communities. These efforts are essential for cultivating long-term relationships that engage and empower individuals, address health disparities, and support equitable, high-quality healthcare delivery.

People-centered care is a transformational approach to healthcare services delivery, emphasizing the needs, preferences, and values of patients, their families, and the communities in which they live, work, and play. By adopting cultural humility and empathy, healthcare delivery systems can foster meaningful relationships between patients and providers, leading to improved health outcomes, greater patient satisfaction, and positive patient expereince. Central to this approach is the active involvement of patients in shared decision-making processes and the unwavering respect for their autonomy, ensuring care is tailored to individual contexts, needs, preferences, and values. Cultural humility demands a continuous commitment to self-reflection and lifelong learning from healthcare providers, enabling them to understand and honor diverse cultural backgrounds, traditions, and experiences. This principle ensures that healthcare services delivery is not only inclusive but also responsive to the unique needs of individuals and communities, particularly those that have historically experienced systemic inequities (Turk, 2020). Empathy further reinforces people-centered care, as it allows providers to connect with patients on an emotional level, demonstrating genuine concern for their health, wellness, well-being, and resiliency while fostering trust in the patient-provider relationship. Shared decision-making is a cornerstone of people-centered care, involving patients as active participants in their life course healthcare journeys. This collaborative process engages and empowers individuals to make informed choices about their healthcare based on a clear understanding of their intervention options, potential outcomes, and personal values. By respecting patient autonomy, healthcare delivery systems world-wide can uphold ethical standards and promote self-determination, ensuring that healthcare services delivery aligns with what matters most to each individual (Beauchamp & Childress, 2019). Communities play a pivotal role in shaping people-centered care by providing insights into an individual's and community's health determinants. Integrating community perspectives allows global healthcare systems to address upstream factors that influence health (i.e., responsible for around 80 % of all health outcomes) and implement empirically-derived interventions that are more effective and equitable (World Health Organization, 2020). For instance, engaging local leaders and organizations can lead to tailored outreach programs that promote health literacy and proactive healthcare services. By focusing on people-centered care with cultural humility and empathy, global healthcare delivery systems can create a clinical framework that respects individual autonomy, fosters trust, and promotes inclusive, equitable practices. This approach not only enhances the quality and safety of care delivered but also supports long-term relationships that engages and empowers individuals and communities to thrive.

These characteristics help ensure that global healthcare delivery systems operate safely, efficiently, effectively, ethically, and sustainably; ultimately leading to a population health management strategy with holistic, integrated and comprehensive primary health care that will ensure better health outcomes for individuals, their families and the communities where they live, work and play.

3.3 Building Capacity into Healthcare Stewardship.

Building capacity in people, processes, and technology related to healthcare services delivery and healthcare stewardship involves a strategic approach to enhance skills, improve complex and adaptive healthcare delivery systems, and leverage information and technological advancements (Anderson et al., 2020; McDarby et al., 2022; Milstein et al, 2020; Nasiri et al., 2021).

Investing in the people who form the foundation of healthcare delivery systems world-wide is critical for achieving environmentally sustainable, equitable, and innovative healthcare services delivery. Training and education should focus on equipping healthcare professionals with the necessary hard and soft skills to uphold healthcare stewardship principles, engage in ethical decision-making, and manage medical supply chains effectively. Comprehensive continuing professional education opportunities are essential for keeping professionals informed about evidence-based best clinical practices, emerging trends in healthcare services delivery, and advancements in healthcare information systems and medical technology, ensuring that they remain adaptable in a rapidly evolving field (Institute of Medicine, 2011). Leadership development plays a pivotal role in fostering strategic and ethical governance within global healthcare delivery systems. Identifying and mentoring potential leaders, coupled with robust value-driven training programs that emphasize critical thinking, logical reasoning, problem solving, transparency, accountability, and integrity, builds a cadre of professionals and healthcare stakeholders prepared to navigate complex healthcare challenges. Hard skills, such as operational management, and soft skills, such as communication and emotional intelligence, are both foundational to effective leadership (Goleman, 2006). Interdisciplinary and multisectoral collaboration is another cornerstone of innovation and progress in healthcare services delivery, encouraging teamwork and communication (e.g., The Learning Healthcare Organization) among research scientists and clinical practitioners to speed up translation of biomedical research into clinical interventions, the classic "bench to bed" phenomenon. Cross-disciplinary workshops, podcasts, and seminars create spaces for collaborative problem-solving and the exchange of creative ideas, addressing the multifaceted challenges of healthcare delivery. Finally, mentorship and professional support systems strengthen the healthcare workforce by fostering knowledge-sharing and continuous learning. Establishing mentorship programs allows experienced professionals to guide less experienced colleagues, while creating learning healthcare systems and communities of practice facilitates the exchange of best practices, resources, and innovations (Wenger, 1998). Together, these initiatives and equity across all healthcare delivery systems worldwide are equipped with skilled, motivated, and forward-thinking professionals, capable of meeting current and future demands.

A robust and systematic process is fundamental for ensuring the efficient and effective delivery of healthcare services while maintaining transparency, accountability, and a commitment to quality improvement science. Standardization is a cornerstone of this process, achieved by developing and implementing operations, clinical policies, protocols, and guidelines based on empirically-driven evidence-based best clinical practices. Regular reviews and updates are essential to align these standards with emerging research findings and advancements in technology, ensuring they remain relevant and effective (Institute of Medicine, 2001). Quality improvement (QI) initiatives further enhance healthcare practices by employing continuous quality improvement (CQI) methodologies to identify, analyze, and refine processes. QI science provides a structured framework that fosters innovation and ensures measurable improvements in healthcare outcomes (Deming, 1986). Performance measurement and monitoring are critical for evaluating the healthcare services delivery processes and driving accountability. Designing and implementing key performance indicators (KPIs) allows organizations to measure success, identify inefficiencies, and track progress over time. Leveraging data analytics enhances these efforts by pinpointing areas for improvement and offering actionable insights. Transparent systems, such as just-in-time dashboards and annual performance scorecards, ensure that stakeholders have access to real-time information, promoting trust with stakeholders and enabling timely decision-making (Porter, 2010). Effective communication is integral to the process, facilitating information flow within and beyond the healthcare delivery system. Developing factual, clear, and accessible communication channels ensures that stakeholders—from clinical teams to administrators and patients—are aligned with organizational goals. Regular meetings, detailed reports, and the use of digital tools enhance connectivity and engagement, fostering a unified understanding of the mission and vision of the healthcare organization (Nutbeam, 2008). By integrating these elements—standardization, quality improvement, performance measurement, and effective communication—healthcare systems can create a transparent, responsive, and adaptable process that consistently delivers high-quality care to diverse populations.

Technology is a critical enabler of modern global healthcare delivery systems, offering innovative tools to improve efficiency, accessibility, and overall healthcare services delivery while safeguarding patient data and privacy. Robust health information systems (HIS) serve as the backbone of healthcare delivery operations, providing platforms to manage patient data, streamline clinical workflows, and enhance evidence-based decision-making. To optimize HIS functionality, organizations must execute comprehensive data-sharing and privacy agreements that ensure the secure exchange of sensitive personal information across multiple systems. Additionally, ensuring interoperability between these systems facilitates seamless data exchange, enabling better coordination of healthcare services delivery and reducing redundant efforts (Institute of Medicine, 2011). Telehealth and remote patient monitoring (RPM) technologies have transformed access to healthcare services, particularly for underserved and marginalized populations. Establishing free broadband internet capabilities is essential for closing the digital divide and ensuring equitable access to telemedicine services. Telehealth platforms extend care to remote areas by enabling virtual consultations and real-time patient support, while remote monitoring tools offer innovative solutions for tracking chronic health conditions and delivering timely interventions to improve patient outcomes (World Health Organization, 2022). The application of data analytics and artificial intelligence (AI) further transforms healthcare services delivery by generating actionable insights into trends, outcomes, and resource utilization. Leveraging advanced AI and machine learning (ML) algorithms empowers healthcare providers to predict patient outcomes, identify inefficiencies, and optimize treatments with unprecedented predictability and speed. These technologies not only enhance operational efficiency but also facilitate more personalized and proactive approaches to care delivery (Jiang et al., 2017). Cybersecurity is foundational to ensuring the protection of patient health information and maintaining trust within global healthcare delivery systems. Implementing robust cybersecurity measures, regularly updating protocols, and conducting annual training on protected health and personally identifiable information protection for all staff members are crucial steps in safeguarding sensitive information and mitigating risks. Proactive cybersecurity strategies ensure that technological advancements in healthcare services delivery are accompanied by unwavering attention to privacy and data security (National Institute of Standards and Technology, 2018). By embracing technology through HIS, telehealth, RPM, data analytics, AI, ML, and cybersecurity, healthcare delivery systems world-wide can create more efficient, inclusive, and resilient systems of health and models of care that meet the needs of diverse populations while advancing the global health landscape.

By prioritizing the development of people, refining processes, and harnessing innovative technology, healthcare organizations and delivery systems world-wide can cultivate a resilient and responsive infrastructure capable of addressing complex and adaptive challenges. Building capacity in these three interconnected areas ensures a holistic and integrated approach to healthcare stewardship, fostering accountability, transparency, and equity across all healthcare delivery systems. Empowering employed staff and clinicians through targeted training programs, leadership development, and interdisciplinary collaboration strengthens the healthcare workforce, fostering an environment of shared problem-solving and innovation. This approach ensures the delivery of high-quality care at the most economical cost, effectively reducing the financial burden on patients. By minimizing medical debt, healthcare delivery systems around the globe can prevent the perpetuation of financial hardship as a legacy for future generations, creating a more equitable and sustainable model of care that prioritizes both immediate health needs and long-term financial well-being for families and communities. Streamlining processes through standardization, continuous quality improvement, performance monitoring, and effective communication optimizes operational efficiency and ensures consistency in care delivery. Leveraging technology, including health information systems, telehealth, artificial intelligence, and cybersecurity, provides innovative tools to expand access, improve resource utilization, and safeguard sensitive data. Collectively, these efforts create a global healthcare ecosystem equipped to achieve sustainable and equitable health outcomes while advancing a culture of excellence, trust, and adaptability for both present and future generations.

3.4 Sustainability in Healthcare Stewardship.

Sustainability in healthcare stewardship, as previously introduced, refers to the responsible allocation, management, and appropriate utilization of limited healthcare resources to ensure the long-term viability of healthcare delivery systems while minimizing environmental impact (Enqvist et al., 2018). This involves implementing practices that balance economic efficiency, social equity, and environmental sustainability. Key aspects of sustainable healthcare stewardship include reducing waste, conserving energy, and promoting the use of sustainable materials. Hospitals and healthcare facilities are increasingly adopting green building standards, such as LEED certification, to reduce their GHG emissions (i.e., Scope 1, 2, and 3) and improve energy efficiency. Additionally, implementing waste reduction programs, such as recycling medical supplies and properly disposing of hazardous materials, plays a critical role in promoting environmental sustainability (Sherman et al., 2020).

Another important facet of sustainability in healthcare is the adoption of sustainable procurement practices. This involves sourcing medical supplies, equipment, and pharmaceuticals from suppliers who adhere to environmentally friendly and socially responsible practices. By prioritizing products that are biodegradable, recyclable, or have a lower environmental impact, healthcare organizations can significantly reduce their carbon footprint. Furthermore, fostering long-term partnerships with medical and pharmaceutical suppliers committed to environmental sustainability can drive innovation in the production of green medical technologies and supplies (Handfield et al., 2020).

Healthcare stewardship fundamentally includes the efficient allocation and utilization of financial resources to safeguard the long-term sustainability of healthcare services. Effective financial stewardship is achieved by adopting value-based healthcare systems and models of care that emphasize delivering high-quality care while maintaining cost control. These models prioritize proactive and preventive care, incentivizing healthcare providers to focus on interventions that yield the most meaningful outcomes for patients rather than rewarding volume-based service delivery (Brinkerhoff et al., 2019a). For instance, value-based approaches often include bundled payments, pay-for-performance programs, and accountable care organizations (ACOs), which shift the focus toward patient-centered care while reducing the financial burden associated with unnecessary medical interventions.

Key features of value-based healthcare systems include optimizing resource utilization, improving care coordination, and managing chronic conditions and reactive care effectively. These efforts play a critical role in minimizing inappropriate hospital readmissions, emergency room visits, and duplicative testing, ultimately leading to significant cost savings and enhanced patient outcomes. Healthcare providers are incentivized to monitor and manage patient health comprehensively, aligning financial goals with better clinical results (World Health Organization, 2022).

Data analytics and health information technology (HIT) further amplify the efficiency of healthcare stewardship by enabling healthcare delivery systems world-wide to source, manage, and allocate resources strategically. Advanced data analytics tools allow healthcare organizations and delivery systems to identify healthcare services delivery trends, forecast patient needs, and pinpoint inefficiencies within processes. These insights support informed decision-making and ensure that financial resources are used in areas with the greatest potential impact (Jiang et al., 2017). Moreover, HIT enhances operational efficiency by facilitating seamless information sharing across healthcare delivery systems, enabling real-time monitoring, and improving care coordination. Ensuring interoperability between HIT platforms strengthens communication among providers and supports continuity of care, further optimizing system performance.

Integrating value-based care models, data-driven strategies, and advanced information and medical technologies creates a robust foundation for environmentally sustainable healthcare services delivery. By focusing on both cost containment and quality improvement, healthcare delivery systems can foster resilience, equity, and long-term viability, meeting the evolving needs of individuals and communities.

Community engagement and education are also essential components of sustainable healthcare stewardship (Mitchell et al., 2022). Educating patients and communities about moving away from reactive chronic condition care to multifaceted proactive care, encouraging healthy lifestyles and behavior, and the responsible use of healthcare services can reduce the overall demand for medical interventions and promote better health outcomes. Community health programs that address health-related social needs (HRSNs) and other determinants of health, such as access to nutritious food, safe housing, and education, are also crucial for fostering long-term health, wellness, well-being, and resilience. By investing in community health and focusing more on proactive care rather than reactive care, healthcare organizations and healthcare delivery systems can create a more environmentally sustainable and resilient global healthcare delivery system.

Lastly, environmental sustainability in healthcare stewardship demands an enduring commitment to innovation, progress, and adaptability (Bornbusch et al., 2014). This entails remaining informed about the latest breakthroughs in medical research, emerging technologies, and best practices that support the sustainability of healthcare services delivery. Healthcare delivery systems around the globe must cultivate a culture of continuous learning and quality improvement, empowering their workforce to integrate environmentally sustainable practices across operations and adopt cutting-edge medical technologies and health information systems designed to enhance efficiency, reduce costs, and minimize environmental impact. This includes leveraging renewable energy solutions, waste reduction strategies, and green design principles in healthcare facilities to mitigate the sector's ecological footprint.

The pursuit of environmental sustainability also involves integrating healthcare delivery systems that are resilient to evolving healthcare demands and adaptable to global challenges, including man-made climate change and resource constraints. By focusing on innovative approaches such as telemedicine, remote patient monitoring, and precision medicine, healthcare systems can deliver patient-centered care while conserving resources and promoting long-term operational efficiency (World Health Organization, 2022). Furthermore, environmental sustainability must be deeply embedded within organizational policies, processes, and practice strategies to ensure alignment with broader public health goals, equity, and the needs of future generations.

By prioritizing environmental sustainability across every dimension of healthcare service delivery—clinical practices, infrastructure, workforce development, and technology—healthcare organizations and delivery systems world-wide can ensure not only the long-term health, wellness, and well-being of patients but also the protection and stewardship of our planet's limited healthcare resources. This holistic approach reaffirms the healthcare sector's role in advancing the public's health and environmental sustainability.

3.5 References

1. Aishammari, Y. F. H., Alharbi, M. N., Alanazi, H. F., et al. (2020). Barriers and facilitators to evidence-based practice adoption among healthcare practitioners. *International Journal of Health & Medical Sciences, 3*(1), 160–166.
2. Anderson, J. E., Ross, A. J., Macrae, C., et al. (2020) Defining adaptive capacity in healthcare: a new framework for researching resilient performance. *Applied Ergonomics* 87: 103111.
3. Atsma, F., Elwyn, G., & Westert, G. (2020). Understanding unwarranted variation in clinical practice: A focus on net-

work effects, reflective medicine, and learning health systems. *International Journal for Quality in Health Care, 32*(4), 271–274.
4. Beauchamp, T. L., & Childress, J. F. (2019). *Principles of biomedical ethics* (8th ed.). Oxford University Press.
5. Bornbusch, A. W., et al. (2014). Strengthening healthcare systems for sustainability: A framework for innovation. *Global Health: Science and Practice, 2*(2), 123–135.
6. Braveman, P., Egerter, S., & Williams, D. R. (2011). The social determinants of health: Coming of age. *Annual Review of Public Health, 32*(1), 381–398.
7. Brinkerhoff, D. W., Cross, H. E., Sharma, S., et al. (2019) Stewardship and health systems strengthening: An overview. *Public Administration and Development* 39(1), 4-10.
8. Bornbusch, A., Dickens, T., Hart, C., et al. (2014). A stewardship approach to shaping the future of public health supply chain systems. *Global Health: Science and Practice* 2(4): 403-409.
9. Christensen, C. M., Grossman, J. H., & Hwang, J. (2009). *The innovator's prescription: A disruptive solution for health care.* McGraw Hill.
10. Deming, W. E. (1986). *Out of the crisis.* MIT Press.
11. Enqvist, J. P., West, S., Masterson, V. A., et al. (2018) Stewardship as a boundary object for sustainability research: Linking care, knowledge and agency. *Landscape and urban planning* 179: 17-37.
12. Fidler, D. P. (2017). *Emerging infectious diseases and global health security: The role of international law.* Global Health Governance, 10(1), 22-40.
13. Goleman, D. (2006). *Emotional intelligence: Why it can matter more than IQ.* Bantam Books.
14. Handfield, R., Finkenstadt, D. J., Schneller, E. S., et al. (2020) A commons for a supply chain in the post-COVID-19 era: the case for a reformed strategic national stockpile. *The Milbank Quarterly* 98(4), 1058-1090.

15. Institute for Healthcare Improvement (IHI). (2021). *Improving health care quality: Tools and techniques*. Institute for Healthcare Improvement: Boston, MA.
16. Institute of Medicine (IOM). (2001). *Crossing the quality chasm: A new health system for the 21st century*. National Academies Press.
17. Institute of Medicine (IOM). (2003). *Health literacy: A prescription to end confusion*. National Academies Press.
18. Institute of Medicine (IOM). (2003). *Ensuring quality through healthcare governance*. National Academies Press.
19. Institute of Medicine (IOM). (2011). *The future of nursing: Leading change, advancing health*. National Academies Press.
20. Institute of Medicine (IOM). (2013). *Best care at lower cost: The path to continuously learning health care in America*. National Academies Press.
21. Jawhari, B., Ludwick, D., Keenan, L., Zakus, D., & Hayward, R. (2016). Benefits and challenges of EMR implementations in low-resource settings: A state-of-the-art review. *BMC Medical Informatics and Decision Making, 16*(116).
22. Jiang, F., et al. (2017). Artificial intelligence in healthcare: Past, present, and future. *Stroke and Vascular Neurology, 2*(4), 230–243.
23. Kaplan, R. S., & Haas, D. A. (2025). Balancing cost containment and quality improvement in healthcare: A strategic framework. *Journal of Healthcare Management, 70.*
24. Khatri, R. B., Endalamaw, A., Erku, D., et al. (2023). Preparedness, impacts, and responses of public health emergencies towards health security: Qualitative synthesis of evidence. *Archives of Public Health, 81*(208).
25. Kickbusch, I., Allen, L., & Franz, C. (2016). The commercial determinants of health. *The Lancet Global Health, 4*(12), e895-e896.
26. Lear, J. L., Fleig-Palmer, M. M., Hodge, K. A., et al. (2016) Business fundamentals for healthcare providers: Ensuring effective practice management and good stewardship. *Journal of Health Administration Education* 33(1): 141-162.

27. McDarby, G., Mustafa, S., Zhang, Y., et al. (2022) Essential Public Health Functions in Ireland: Perspectives to strengthen capacities and stewardship. *European Journal of Public Health*, 32(3): 129-207.
28. Milstein, B., Erickson, J., Gouveia, T., et al. (2020) Amplifying stewardship: characteristics and trends stewards consider when expanding equitable well-being. RWJF: The Ripple Foundation.
29. Mitchell, J., Cooke, P., Ahorlu, C., et al. (2022) Community engagement: The key to tackling Antimicrobial Resistance (AMR) across a One Health context? *Global Public Health*, 17(11): 2647-2664.
30. Nasiri, T., Yazdani, S., Shams, L., et al. (2021) Stewardship of noncommunicable diseases in Iran: a qualitative study. *International Journal of Health Governance* 26(2): 179-198.
31. National Academies of Sciences, Engineering, and Medicine. (2018). *Crossing the global quality chasm: Improving health care worldwide*. Washington, DC: The National Academies Press.
32. National Academies of Sciences, Engineering, and Medicine. (2025). *Enhancing the resilience of health care and public health critical infrastructure: Proceedings of a workshop–in brief.* Washington, DC: The National Academies Press.
33. National Institute of Standards and Technology. (2018). *Cybersecurity framework*. National Institute of Standards and Technology.
34. Nutbeam, D. (2008). The evolving concept of health literacy. *Social Science & Medicine, 67*(12), 2072–2078.
35. O'Sullivan, L., Aldasoro, E., O'Brien, Á., et al. (2022). Ethical values and principles to guide the fair allocation of resources in response to a pandemic: A rapid systematic review. *BMC Medical Ethics, 23*(70).
36. Office of the National Coordinator for Health Information Technology (ONC). (2021). *Improving interoperability in healthcare systems*. Office of the National Coordinator for Health Information Technology.

37. Pichler, P.-P., Jaccard, I. S., Weisz, U., & Weisz, H. (2019). International comparison of health care carbon footprints. *Environmental Research Letters, 14*(6), 064004.
38. Porter, M. E. (2010). What is value in health care? *New England Journal of Medicine, 363*(26), 2477–2481.
39. Religioni, U., Barrios-Rodríguez, R., Requena, P., et al. (2025). Enhancing therapy adherence: Impact on clinical outcomes, healthcare costs, and patient quality of life. *Medicina, 61*(1), 153.
40. Rosen, M. A., et al. (2018). Teamwork in healthcare: Key discoveries enabling safer, high-quality care. *American Psychologist, 73*(4), 433–450.
41. Sherman, J. D., Thiel, C., MacNeill, A., et al. (2020) The green print: advancement of environmental sustainability in healthcare. *Resources, Conservation and Recycling* 161: 104882.
42. Teixeira de Siqueira Filha, N., Li, J., Phillips-Howard, P. A., et al. (2022). The economics of healthcare access: A scoping review on the economic impact of healthcare access for vulnerable urban populations in low- and middle-income countries. *International Journal for Equity in Health, 21*(191).
43. Turk, M. T. (2020). Empathy and cultural humility in healthcare communication: Foundations for patient-centered care. *Journal of Patient Experience, 7*(4), 482–488.
44. United States Government Accountability Office (GAO). (2023). *Why health care is harder to access in rural America.* United States Government Accountability Office.
45. United States Department of Health and Human Services (HHS). (2023). *Healthcare cybersecurity: Keeping patient data safe.* United States Department of Health and Human Services.
46. Wenger, E. (1998). *Communities of practice: Learning, meaning, and identity.* Cambridge University Press.
47. World Health Organization (WHO). (2020). *Healthcare-associated infection prevention and control.* World Health Organization.
48. World Health Organization (WHO). (2020). *Managing epidemic threats across borders.* World Health Organization.

49. World Health Organization (WHO). (2020). *Health equity and literacy: Addressing disparities*. World Health Organization.
50. World Health Organization (WHO). (2020). *Health equity and social determinants*. World Health Organization.
51. World Health Organization (WHO). (2020). *Social determinants and people-centered health systems*. World Health Organization.
52. World Health Organization (WHO). (2021). *Year of Health and Care Workers 2021: Facts on health and care workers*. World Health Organization.
53. World Health Organization (WHO). (2021). *Making choices in health: WHO guide to cost-effectiveness analysis*. World Health Organization.
54. World Health Organization (WHO). (2021). *Building capacity for health system resilience*.
55. World Health Organization (WHO). (2022). *Equity in health and universal health coverage*. World Health Organization.
56. World Health Organization (WHO). (2022). *Global strategy on digital health*. World Health Organization.
57. World Health Organization (WHO). (2022). *Healthcare and sustainability: A global perspective*. World Health Organization.
58. World Health Organization (WHO). (2022). *Value-based care and the future of sustainable health systems*. World Health Organization.

4.0

A Brief Overview of the Global Healthcare Delivery System.

Including a brief overview of the global healthcare delivery system in a book on healthcare stewardship is essential for providing context and framing the complexities of healthcare service delivery across diverse regions and populations. Healthcare stewardship emphasizes equity, sustainability, and resilience, and to address these principles effectively, readers need to understand the varying structures, challenges, and opportunities within global healthcare systems.

Such an overview sets the stage for exploring how healthcare systems differ in governance, resource allocation, accessibility, and outcomes, highlighting the health disparities that stewardship aims to mitigate. It provides readers with a foundation to appreciate the inextricable interconnectedness of healthcare services delivery across borders, especially in light of transnational challenges such as infectious disease outbreaks, climate-related health threats, and refugee migration.

Furthermore, understanding the global landscape facilitates discussions on empirically-derived best clinical practices, innovations, and collaborative strategies that can be applied to strengthen healthcare systems globally. It also helps readers grasp how healthcare stewardship principles can be adapted to diverse contexts, whether in high-income countries (HICs) with advanced infrastructures or low- and middle-income countries (LMICs) facing systemic challenges. This foundation enhances the relevance of stewardship and reinforces its universal importance in fostering environmentally sustainable healthcare solutions for current and future generations.

4.1 The Global Healthcare Delivery System.

The global healthcare delivery system is a complex and interconnected network of structures, policies, and practices that work to provide healthcare services to diverse populations across the world. It encompasses a wide range of healthcare delivery models, governance frameworks, funding mechanisms, and resource allocation strategies, all of which vary significantly by country and region. Understanding the global healthcare delivery system is essential for addressing pressing global health challenges, fostering equity, and promoting the principles of healthcare stewardship.

Healthcare delivery systems worldwide can be broadly categorized into public, private, or mixed models, each with unique strengths and limitations. In high-income countries, such as those in Europe and North America, healthcare delivery systems often benefit from advanced infrastructure, comprehensive health insurance schemes, and a strong focus on quality and innovation. Universal health coverage (UHC) is a cornerstone of many of these HICs systems, emphasizing equitable access to healthcare services without financial hardship (World Health Organization, 2022). In contrast, low- and middle-income countries (LMICs) face significant challenges, including underfunded healthcare delivery systems, workforce shortages, and inequities in access to care. These disparities underscore the critical importance of global cooperation and knowledge-sharing to ensure that all healthcare delivery systems anywhere in the world are equipped to meet the needs, preferences, and values of their populations.

The interconnectedness of global healthcare delivery systems becomes evident in the face of transnational health challenges, such as pandemics, climate-related health crises, and the growing burden of noncommunicable diseases (NCDs). Effective coordination and collaboration among countries, international health-related organizations, and healthcare providers are essential to address these challenges and strengthen healthcare system resilience. For instance, initiatives like the International Health Regulations (IHR) facilitate collective action to prevent and respond to world-wide health emergencies, underscoring the need for coordinated approaches to healthcare services delivery (Fidler, 2017).

Another defining feature of the global healthcare delivery system is its engagement with health-related social needs (HRSNs) like safe and stable housing, food security, reliable transportation and other determinants of health, such as education, income, and living conditions. Addressing these upstream health determinants is essential for promoting health equity and environmental sustainability, as they significantly influence health outcomes. In recent years, the integration of digital health technologies, such as telemedicine and electronic health records (EHRs), has emerged as a transformational force in global healthcare services delivery, bridging gaps in access and improving efficiency across diverse settings (World Bank, 2022).

In summary, the global healthcare delivery system is an intricate and dynamic framework shaped by the interplay of governance, resources, technology, and societal needs. It encompasses diverse models and approaches to healthcare delivery, reflecting the unique historical, economic, and cultural contexts of different nations. Governance plays a pivotal role in ensuring that healthcare delivery systems are accessible, equitable, and efficient. Effective governance involves strategic policy frameworks, oversight mechanisms, coalition-building, and accountability measures (WHO, 2025). These elements are essential for addressing disparities and promoting social justice within healthcare systems.

Resources, including healthcare infrastructure and workforce, are fundamental to the system's functionality. A robust infrastructure—comprising hospitals, clinics, and community health centers—paired with a skilled workforce, ensures the delivery of quality care. However, disparities in resource allocation often lead to inequities in healthcare services access and outcomes, particularly in underfunded regions.

Technology serves as a transformational force within the global healthcare delivering system. Innovations in advanced medical technology, telemedicine, and data analytics have revolutionized healthcare services delivery, enabling more efficient and personalized care. Yet, the integration of technology also presents challenges, such as ensuring equitable access and addressing ethical concerns related to data privacy and security.

Societal needs and values further shape global healthcare delivering systems, influencing priorities and approaches to care. For instance, some nations prioritize publicly funded healthcare services to promote universal access, while others rely on private insurance and market-driven solutions. These variations underscore the complexity and diversity of the global healthcare landscape.

Recognizing this diversity and complexity is crucial for advancing healthcare stewardship. Healthcare stewardship emphasizes environmentally sustainable and equitable healthcare services delivery, advocating for collaboration among stakeholders to address systemic challenges. By fostering partnerships and leveraging shared resources, healthcare delivery systems world-wide can work towards achieving universal health coverage (UHC) and improving health outcomes worldwide (Debie et al., 2022)

The global healthcare delivery system is a multifaceted and evolving entity that requires a nuanced understanding of governance, resources, technology, and societal needs. Addressing its challenges and leveraging its opportunities are essential for building resilient and equitable healthcare systems that meet the needs of all populations (Debie et al., 2022).

4.2 Comparing High-Income Countries (HICs) to Low-and Middle-Income Countries (LMICs) Healthcare Delivery Systems.

High-income countries (HICs) and low- and middle-income countries (LMICs) exhibit distinct characteristics in their healthcare delivery systems and services, shaped by economic, social, and infrastructural factors. These differences highlight the challenges and opportunities for improving global health equity and collaboration (Mounier-Jack et al., 2017; NGDI, 2023).

1. *Governance and Policy*: In HICs, countries like the United Kingdom have implemented universal healthcare systems such as the National Health Service (NHS), which ensures access to healthcare for all citizens through public funding. In contrast, LMICs like Nigeria often rely on a mix of public and private healthcare providers, with significant dependence on international aid and donor-funded programs. This reliance can lead to fragmented governance and challenges in long-term sustainability.
2. *Infrastructure*: HICs benefit from advanced healthcare infrastructure. For example, Germany's healthcare delivery system includes state-of-the-art hospitals and diagnostic facilities that support efficient and high-quality care. On the other hand, LMICs like Uganda face challenges with inadequate infrastructure, particularly in rural areas, where access to essential healthcare services is limited.
3. *Technology Integration*: HICs lead in integrating technology into their healthcare delivery systems. For instance, the United States has widely adopted electronic health records (EHRs), health information exchanges (HIEs), and telemedicine platforms, enhancing efficiency and patient outcomes. In LMICs, countries like India have made strides in telemedicine to reach underserved populations, but barriers such as limited broadband internet access and affordability persist.

4. *Access and Equity*: HICs, such as Sweden, provide equitable access to healthcare services through universal health coverage (UHC) and robust social safety nets. In LMICs, countries like Bangladesh face significant disparities in access due to socioeconomic factors and geographic challenges, particularly in remote areas.
5. *Workforce*: HICs have a higher density of healthcare professionals. For example, Japan boasts a strong healthcare workforce with excellent patient-to-provider ratios. In contrast, LMICs like Ethiopia experience shortages of trained healthcare workers, exacerbated by brain drain and limited educational opportunities.
6. *Disease Burden*: HICs focus on managing chronic diseases and aging populations. For instance, Canada has comprehensive programs for managing diabetes and cardiovascular diseases. LMICs, such as Kenya, prioritize combating infectious diseases like malaria and tuberculosis while also addressing the growing burden of non-communicable diseases.

Both HICs and LMICs face challenges in establishing an integrated system of health and addressing the growing burden of chronic diseases. For example, while HICs like the United States grapple with rising healthcare costs and disparities among marginalized populations, LMICs like Haiti contend with vertical programs that disrupt coordinated healthcare service delivery.

Learning from each other's experiences can foster innovation and improve healthcare delivery systems globally. For instance, LMICs like Rwanda have developed cost-effective solutions, such as community health worker programs, that HICs can adapt to address similar challenges. Conversely, HICs can support LMICs through technology transfer, capacity building, and policy development.

4.3 Primary Health Care: Cornerstone of Global Healthcare Delivery Systems.

The Primary Health Care (PHC) model is employed by many global healthcare delivery systems because it is a proven strategy for achieving universal health coverage (UHC) and addressing health inequities. It is grounded in the principles of equity, solidarity, and social justice, as outlined in the 1978 Declaration of Alma-Ata and reaffirmed in the 2018 Astana Declaration (WHO & UNICEF, 2022). By focusing upstream on the health-related social needs (HRSNs) and the broader determinants of health (i.e., economic, environmental and commercial), such as stable and safe housing, food security, viable transportation options, and job training with employment opportunities, the PHC model ensures that healthcare delivery systems are not only providing reactive care (e.g., chronic disease with care coordination) but also proactive in its approach to health, wellness, well-being, and resilience (e.g., health prevention, health promotion, health protection, early warning disease surveillance with population health management, and emergency preparedness, resilience and rapid response).

Moreover, the PHC model is cost-effective. Studies have shown that investing in PHC can lead to significant reductions in healthcare costs by preventing diseases and reducing the need for expensive and inappropriate hospital care. It also strengthens health systems' resilience, enabling them to respond effectively to emergencies, such as pandemics (WHO, 2023).

The PHC model is most effective when implemented through a whole-of-society approach that integrates its three core components:

a) *Multisectoral Policy and Action*: Addressing the determinants of health requires a comprehensive and collaborative approach across multiple sectors and disciplines, as health outcomes are deeply influenced by factors outside the traditional healthcare delivery system. These determinants include economic stability, education, social and physical environments, and access to essential resources like water and sanitation. Effective multisectoral and interdisciplinary collaboration recognizes that improving health cannot be achieved solely within clinical settings but demands joint efforts from various sectors, such as housing, transportation, education, and agriculture. For instance, improving access to clean water and sanitation is a vital component of this approach, as it directly impacts the prevalence of waterborne infectious diseases. Diseases like cholera and diarrhea, which disproportionately affect vulnerable populations, can be significantly reduced through infrastructure improvements, public awareness campaigns, and policies that prioritize water safety standards. Governments, NGOs, and private sector actors must work together to develop sustainable systems for water purification, wastewater management, and equitable distribution of these resources. Another example of multisectoral collaboration is the implementation of healthy public policies to foster positive behaviors and lifestyles. Anti-smoking campaigns, for instance, exemplify how coordinated efforts from health departments, education ministries, and media outlets can effectively reduce the prevalence of smoking-related illnesses. These campaigns often include measures such as educational programs in schools, public service advertisements, increased taxation on tobacco products, and restrictions on tobacco advertising. Together, these initiatives create a supportive environment that encourages individuals to adopt healthier choices and behaviors. By addressing these determinants of health through cross-sector collaboration, governments and organizations not only improve individual and community health but also strengthen the overall resilience of healthcare delivery systems. Such initiatives ensure both a proactive and reactive approach to patient care using empirically-driven evidence-based clinical practice guidelines, reducing the burden of disease and promoting wellness and well-being for all.

b) *Engaged and Empowered Communities*: Involving communities in healthcare service decision-making fosters a collaborative dynamic between individuals and their primary care providers, ensuring that the care delivered is personalized to meet the unique needs, preferences, and values of each patient. This participatory approach acknowledges the critical role that patients play as partners in their own life course health journey, thereby enhancing the relevance, acceptability, and effectiveness of healthcare interventions. Community health workers (CHWs) and peer navigators are integral to this process, serving as essential conduits between healthcare systems and the communities they aim to serve. CHWs, often drawn from the very communities they represent, possess a deep understanding of the cultural, social, and economic contexts that influence health behaviors and decisions. Their ability to communicate in culturally appropriate ways fosters trust and reduces barriers to accessing healthcare services. Similarly, peer navigators, who share lived experiences with those they support, are uniquely positioned to guide patients through complex healthcare delivery systems, providing both practical assistance and emotional encouragement. By bridging the gap between healthcare providers and patients, CHWs and peer navigators promote equitable access to services, ensuring that marginalized or underserved populations can access the care they need. Their advocacy extends beyond individual interactions to include community-level engagement, where they amplify voices and drive grassroots solutions to health challenges. In doing so, they empower communities to become active participants in shaping health priorities, influencing policy, and co-designing interventions that reflect their collective needs. Empowered communities—those equipped with the knowledge, resources, and agency to engage in health decision-making—are better positioned to identify their own health priorities and advocate for meaningful, sustainable solutions. This empowerment not only improves immediate health outcomes but also strengthens the resilience of communities, enabling them to address systemic health inequities and build a healthier future.

c) *Integrated Health Services*: The Primary Health Care (PHC) model places a strong emphasis on the seamless integration of services across the entire continuum of care. This approach encompasses a comprehensive range of health-related interventions, starting from health promotion, health protection, and health prevention to treatment, rehabilitation, and palliative care. By bridging these elements, the PHC model ensures that individuals receive coordinated, integrated and holistic care, tailored to their specific needs, preferences, and health circumstances throughout their life course health journey.

    a) *Health Promotion*: At its core, the Primary Health Care (PHC) model emphasizes proactive health promotion as a cornerstone of achieving equitable and sustainable health outcomes. This focus empowers individuals and communities to take ownership of their health, fostering a sense of agency and responsibility for personal wellness and well-being. Health promotion activities encompass a wide range of initiatives, such as public health education programs designed to increase awareness about common health risks, preventive measures, and available healthcare resources. For example, workshops on managing chronic conditions or understanding the importance of immunizations equip individuals with actionable knowledge. Community outreach programs further amplify these efforts by directly engaging underserved populations, delivering culturally relevant information, and addressing barriers to care. Campaigns targeting lifestyle factors—such as encouraging balanced nutrition, regular physical activity, and stress management—serve as additional pillars of this approach, enabling communities to build healthier routines that reduce the risk of chronic and preventable diseases. Beyond reducing the immediate burden of disease, health promotion within the PHC model fosters long-term benefits by cultivating a culture of health literacy. This emphasis on education and empowerment equips individuals to make informed decisions about their health, understand when to seek medical care, and actively participate in treatment planning with healthcare providers. By integrating health promotion into the broader continuum of care, the PHC model addresses health challenges at their root, preventing escalation into costly or complex medical interventions. It also reinforces the model's holistic vision of health, which extends beyond the absence of illness to encompass physical, mental, and social well-being. Together, these efforts strengthen the resilience of communities, improve quality of life, and contribute to the realization of universal health coverage (UHC).

b) *Health Protection*: Health protection within the Primary Health Care (PHC) framework is a proactive and multifaceted approach to safeguarding communities from health risks. It focuses on reducing exposure to environmental, biological, and health-related social needs (HRSNs) through strategies such as immunization programs, infectious disease control, and environmental health measures. Initiatives like vaccination campaigns protect populations from preventable diseases, while regulations on water safety, food hygiene, and industrial emissions reduce exposure to harmful environmental factors. These measures collectively strengthen the resilience of healthcare systems and communities, particularly in underserved regions. In addition to addressing immediate health risks, health protection under PHC emphasizes preparedness for emergencies and occupational safety. Policies and practices such as disaster contingency planning, workplace hazard mitigation, and tobacco control regulations ensure populations are shielded from systemic and situational threats. The overarching goal is to create equitable and sustainable systems that minimize harm and promote safety, contributing to universal health coverage (UHC) and better health outcomes globally.

c) *Health Prevention*: Preventing the onset and spread of diseases is a critical component of the Primary Health Care (PHC) model, seamlessly integrated alongside health promotion and protection. Vaccination programs serve as a cornerstone of this approach, targeting preventable diseases such as measles, polio, and influenza. These programs help build herd immunity (threshold of 95 % individuals with vaccination coverage) within communities, significantly reducing the incidence of outbreaks and protecting vulnerable populations. Screening initiatives further bolster disease prevention by enabling early detection of conditions like hypertension, diabetes, or certain types of cancer. Early diagnosis allows for timely intervention, minimizing complications and promoting better health outcomes. Together, these preventive measures reduce the physical and emotional burden of illnesses on individuals while alleviating pressure on healthcare systems. In addition to individual-focused interventions, the PHC model addresses broader environmental risk factors that contribute to disease. Strategies such as improving access to clean water and sanitation, reducing exposure to air and water pollution, and promoting healthier living conditions are essential to mitigating health hazards. These efforts target the root causes of disease, fostering healthier communities and environments. By proactively preventing illnesses and addressing systemic risks, the PHC model strengthens health resilience, reduces healthcare costs, and creates a foundation for sustainable, equitable healthcare systems globally. Its comprehensive approach ensures that prevention is not an isolated action but an integral part of achieving universal health coverage and improving quality of life for all.

d) *Treatment*: The PHC framework ensures that when individuals require medical intervention, the care they receive is comprehensive and person-centered. Primary care providers act as the first point of contact, diagnosing and managing a wide array of health conditions while coordinating with specialists as necessary. This integrated system avoids fragmentation of care and maintains continuity, ensuring that patients feel supported throughout their treatment journey.
e) *Rehabilitation and Palliative Care*: Beyond treatment, the PHC model extends to rehabilitation and palliative services, ensuring that patients regain functionality after acute health events or receive compassionate end-of-life care. Rehabilitation programs, such as physical therapy after injury or illness, are vital for restoring individuals' independence and quality of life. Meanwhile, palliative care focuses on managing chronic conditions and providing emotional and psychological support, affirming the PHC model's commitment to a patient-centric approach.

By integrating these components within a cohesive framework, the PHC model not only improves individual health outcomes but also strengthens the overall efficiency and resilience of healthcare delivery systems. The model's focus on continuity and collaboration ensures that care is not fragmented or isolated but instead follows individuals across their lifespans, adapting to their evolving needs. Ultimately, this integration embodies the core philosophy of PHC: delivering equitable, accessible, and sustainable care that places individuals at the heart of the healthcare experience (WHO, 2023).

The primary purpose of the PHC model is to achieve universal health coverage (UHC) and improve health equity. By bringing health services closer to communities, the model ensures that care is accessible, affordable, and centered on people's needs, preference, and values. It also supports the Sustainable Development Goals (SDGs), particularly Goal 3, which aims to ensure healthy lives and promote well-being for all at all ages (WHO-UNICEF, 2022). Additionally, the PHC model fosters a shift from disease-focused care to a more holistic approach that prioritizes proactive care. This not only improves individual and community health outcomes but also contributes to the overall development of societies.

For the Primary Health Care (PHC) model to achieve success, it requires the seamless integration of advanced primary care services, essential public health functions, and the contributions of community-based organizations (CBOs). These components must operate collaboratively within a person-centered medical home (PCMH) model, which serves as a hub for coordinated, comprehensive, and patient-focused care.

The PCMH model places individuals and their families at the center of healthcare delivery, ensuring that care is tailored to their unique needs, preferences, and values. Advanced primary care services play a crucial role in this framework by offering accessible, high-quality, and holistic care across the continuum of health needs. These services include preventive care, chronic disease management, and acute care, delivered by multidisciplinary teams that emphasize continuity and personalized care. At the same time, essential public health services—such as health surveillance, disease prevention programs, and health promotion campaigns—complement primary care by addressing population-level health challenges and reducing systemic barriers to wellness.

Community-based organizations (CBOs) bring an additional layer of support, acting as trusted intermediaries between healthcare delivery systems and the populations they serve. CBOs offer culturally sensitive services, advocate for community health priorities, and address upstream determinants of health, such as housing, food security, and transportation. When fully integrated into the PCMH model, CBOs enhance care coordination and expand the reach of healthcare services delivery, ensuring that vulnerable populations receive equitable access to care. Together, the alignment of advanced primary care, public health services, and community-based organizations under the PCMH model creates a holistic and resilient healthcare services delivery framework. This integration not only improves individual and population health outcomes but also fosters a more equitable, efficient, and sustainable health system. The World Health Organization (2023) emphasizes that collaborative, person-centered care models like PCMH are critical for realizing the broader goals of the PHC model and achieving universal health coverage.

4.4 Key Performance Indicators for Global Healthcare Delivery Systems.

Key Performance Indicators (KPIs) are measurable values that all organizations use to evaluate their success in achieving specific objectives. In the context of healthcare, KPIs are essential for monitoring, assessing, and improving the quality, efficiency, and equity of healthcare delivery. They provide a framework for tracking progress, identifying areas for improvement, and ensuring accountability in healthcare systems (WHO, 2022; WHO, n.d.; Fallahnezhad et al., 2024).

1. **Access to Healthcare**:
    a) **Definition**: Access KPIs measure the availability and reach of healthcare services to populations.
    b) **Examples**:
        i) Percentage of the population with health insurance coverage.
        ii) Average waiting time for primary care appointments.
        iii) Geographic distribution of healthcare facilities.
    c) **Significance**: These indicators highlight disparities in healthcare access and guide resource allocation to underserved areas.

2. **Quality of Care**:
    a) **Definition**: Quality KPIs evaluate the standard of care provided to patients.
    b) **Examples**:
        i) Patient satisfaction scores.
        ii) Hospital readmission rates within 30 days of discharge.
        iii) Adherence to clinical guidelines for chronic disease management.
    c) **Significance**: They ensure that healthcare services meet established standards and improve patient outcomes.

3. **Efficiency**:
    a) **Definition**: Efficiency KPIs assess the optimal use of limited resources in delivering healthcare (i.e., healthcare stewardship).
    b) **Examples**:
        i) Average length of hospital stay.
        ii) Total cost per patient treated.
        iii) Utilization rates of healthcare resources (e.g., hospital beds, medical equipment).
    c) **Significance**: These indicators help reduce waste and improve the cost-effectiveness of healthcare systems.

4. **Equity**:
    a) **Definition**: Equity KPIs measure the fairness of healthcare delivery across different population groups.
    b) **Examples**:
        i) Disparities in health outcomes across socioeconomic groups.
        ii) Proportion of healthcare services provided to underserved populations.
        iii) Gender parity in access to healthcare services.
    c) **Significance**: They address systemic inequalities and promote inclusive healthy public policies.

5. **Health Outcomes**:
    a) **Definition**: Health outcome KPIs track the overall health status of populations.
    b) **Examples**:
        i) Infant and maternal mortality rates.
        ii) Life expectancy at birth.
        iii) Prevalence and incidence rates of major diseases (e.g., diabetes, cardiovascular diseases).
    c) **Significance**: These indicators reflect the effectiveness of healthcare interventions and healthy public policies.

6. **Sustainability and Resilience**:
    a) **Definition**: These KPIs evaluate the capacity of healthcare delivery systems to withstand and adapt to challenges and adversity.
    b) **Examples**:
        i) Availability of essential medicines and supplies during emergencies.
        ii) Proportion of healthcare facilities with renewable energy sources.
        iii) Preparedness for public health emergencies.
    c) **Significance**: They ensure long-term viability and adaptability of healthcare systems.

7. **Workforce Metrics**:
    a) **Definition**: Workforce KPIs assess the availability and performance of healthcare professionals.
    b) **Examples**:
        i) Ratio of healthcare workers to population.
        ii) Staff turnover rates in healthcare facilities.
        iii) Training and professional development opportunities for healthcare workers.
    c) **Significance**: These indicators support workforce planning and capacity building.

Key Performance Indicators (KPIs) are indispensable tools for evaluating and enhancing global healthcare delivery systems. They provide measurable benchmarks that inform policy decisions, drive improvements, and ensure accountability across diverse aspects of healthcare, including access, quality, equity, efficiency, effectiveness, sustainability, and workforce performance. This framework of evaluation empowers stakeholders—governments, organizations, and providers—to identify systemic challenges, track progress, and implement value-driven and empirically-derived evidence-based interventions.

This list of KPIs represents only a sampling of the numerous metrics utilized across global healthcare delivery systems. These indicators vary widely depending on regional priorities, healthcare structures, and societal contexts. As healthcare systems continue to evolve, the adoption of KPIs will play an increasingly pivotal role in addressing global challenges, such as achieving Universal Health Coverage (UHC) and meeting the Sustainable Development Goals (SDGs).

The overall significance of KPIs extends beyond their immediate applications in performance measurement. They foster transparency and trust among patients and stakeholders, facilitate resource optimization, and promote equity by addressing disparities in healthcare access and outcomes. By providing a standardized framework for monitoring and assessment, KPIs enable healthcare systems to operate more effectively and sustainably with resiliency, ensuring better health for populations worldwide. Their importance cannot be overstated in the context of advancing global healthcare delivery toward a model that prioritizes both human wellness and well-being and system-wide efficiency and equity.

4.5 Measuring the Unintended Consequences of Global Healthcare Delivery Systems.

The unintended consequences of healthcare services delivery are an important area of study as they can significantly impact patient outcomes, healthcare costs, and system efficiency (Berry et al., 2022; Norton et al., 2020; Rahman et al., 2020).

One common unintended consequence is medical errors, which can occur due to miscommunication, misdiagnosis, or procedural mistakes. Quantifying medical errors involves analyzing data from incident reports, patient records, and malpractice claims. Studies estimate that medical errors are the third leading cause of death in the US (Ahmad et al., 2021), highlighting the critical need for robust performance tracking systems to track and mitigate these errors. By using metrics such as error rates, patient harm incidents, and associated costs, healthcare organizations and delivery systems can better understand the prevalence and impact of medical errors and implement strategies to reduce them (Vahidi et al., 2020).

Another significant unintended consequence is the issue of healthcare-associated infections (HAIs). These infections can occur in various healthcare settings, including hospitals and long-term care facilities, and can lead to severe adverse patient outcomes and increased healthcare costs. Quantification of HAIs involves tracking infection rates through disease surveillance systems, analyzing patient data, and monitoring the effectiveness of infection control measures. Metrics such as the rate of specific infections per 1,000 patient days and the cost of treatment for these infections provide valuable insights. By quantifying HAIs, healthcare organizations can identify high-risk areas, implement targeted interventions, and monitor the effectiveness of infection prevention strategies (Haque et al., 2020).

The overuse, underuse, and/or misuse of diagnostics and treatments is another unintended consequence that can be quantified (Mekonnen et al., 2020). This includes unnecessary diagnostic tests, inappropriate prescribing of medications, and inappropriate treatment of acute and chronic conditions, all of which can lead to patient harm and increased costs. Quantifying inappropriate use involves analyzing healthcare utilization data, reviewing patient records, and comparing practices against established empirically-derived evidence-based clinical practice guidelines. Metrics such as the frequency of specific tests or treatments per patient population and the associated costs of inappropriate use provide a clear picture of the extent of the problem. By recognizing patterns of inappropriate care, healthcare providers can focus on implementing evidence-based clinical practices, enhancing patient safety, and minimizing excessive healthcare expenditures.

The quantification of patient readmissions is also foundational in understanding unintended consequences (Becker et al., 2021). Readmissions often indicate inadequate initial treatment or poor post-discharge care planning and can be costly for both patients and healthcare delivery systems. Tracking readmission rates, especially for conditions like heart failure, pneumonia, and chronic obstructive pulmonary disease (COPD), can highlight gaps in transitions of care and post-discharge support. Metrics such as the 30-day readmission rate and the cost of readmissions help quantify this issue. By analyzing these metrics, healthcare providers can implement interventions such as improved discharge planning, patient and care-giver education, and follow-up care to reduce readmissions and enhance patient outcomes.

The unintended consequences of healthcare delivery on behavioral health—encompassing mental health illnesses and substance use disorders—represent a critical area that necessitates robust quantification. The pressures inherent to healthcare systems can impose significant emotional and psychological burdens on both patients and providers, contributing to a rise in the so-called diseases of despair, which include conditions such as depression, anxiety, and suicide. This phenomenon is exacerbated by systemic stressors, such as insufficient support structures, high-stakes decision-making, and the challenges of navigating complex care environments. A particularly alarming trend is the dramatic surge in substance use disorders, especially those linked to synthetic opioids, over the past five years. This crisis has reached global proportions, overwhelming healthcare systems and intensifying the strain on resources. The synthetic opioid epidemic exemplifies the broader behavioral health challenges that demand focused attention in healthcare delivery settings. Quantifying the impact of healthcare delivery on behavioral health requires comprehensive approaches, such as deploying surveys, conducting in-depth interviews, and utilizing validated measurement tools to capture behavioral health outcomes. These methods enable the identification of critical patterns in both incidence and prevalence, shedding light on the systemic factors contributing to mental health and substance use challenges. By understanding these dimensions, healthcare delivery systems can proactively develop support mechanisms to address the unique needs of patients and staff. Such strategies might include establishing mental health and wellness programs, fostering healthier workplace environments, and implementing evidence-based interventions to mitigate risks. Furthermore, addressing these behavioral health dimensions is essential not only for improving individual well-being but also for ensuring the resilience and sustainability of healthcare systems globally (Coombs et al., 2021).

Quantifying the unintended consequences of healthcare delivery involves identifying key performance indicators (KPIs) in areas such as medical errors, healthcare-associated infections, overuse of medical interventions, patient readmissions, and adverse behavioral health impacts. By asking the right questions to obtain appropriate KPI data and then, analyzing the data that is available, healthcare delivery systems can gain a comprehensive understanding of all of their KPIs, implement targeted interventions, and ultimately improve the quality and safety of healthcare delivery.

4.6 The Global Medical Supply Chain

The global medical supply chain is a critical component of healthcare stewardship, ensuring that medical supplies and equipment are available where and when they are needed while minimizing waste and inefficiencies. Efficient supply chain management involves the careful planning, procurement, storage, and distribution of medical products. By leveraging advanced technologies and data analytics, healthcare organizations can forecast demand more accurately, reducing the risk of overstocking or stockouts. This precision in inventory management not only optimizes resource use but also ensures that healthcare providers have access to essential supplies, enhancing patient care and safety (Ziat et al., 2020; Blanchard, 2021)

Additionally, the global medical supply chain plays a pivotal role in cost containment, a key aspect of healthcare stewardship (Ketokivi et al., 2020). By negotiating better prices with suppliers, adopting group purchasing strategies, and implementing value-based procurement practices, healthcare organizations can significantly reduce their expenditure on medical supplies. This cost savings can be redirected toward other critical areas of patient care and facility improvement. Moreover, efficient supply chain operations can reduce operational costs related to storage, transportation, and handling of supplies, contributing to the overall financial sustainability of healthcare institutions.

Environmental sustainability is another crucial area where the medical supply chain impacts healthcare stewardship (AlJaberi et al., 2020; Zhu et al., 2020). By incorporating environmentally friendly practices, such as reducing packaging waste, choosing sustainable materials, and implementing recycling programs, healthcare organizations can lessen their environmental footprint. Sustainable supply chain practices not only benefit the environment but also resonate with the growing emphasis on corporate social responsibility in the healthcare sector. This alignment with sustainability goals can enhance the reputation of healthcare institutions and contribute to broader public health objectives.

Furthermore, the resilience and agility of the global medical supply chain are vital for effective healthcare stewardship, especially during crises such as pandemics or natural disasters (Rehman et al., 2022). A well-prepared and flexible supply chain can quickly adapt to sudden changes in demand, ensuring the continuous availability of critical medical supplies. This resilience is achieved through strategic stockpiling, diversifying suppliers, and utilizing robust logistics networks. By maintaining a reliable supply chain, healthcare organizations can provide uninterrupted care, prevent shortages of essential items, and support public health preparedness for and responses during emergencies (Patel et al., 2022)

Lastly, the global medical supply chain contributes to healthcare stewardship through the adoption of innovative technologies and practices (Yang et al., 2021). Implementing automated inventory systems, utilizing blockchain for enhanced transparency and traceability, and employing artificial intelligence for predictive analytics are just a few examples of how technology can optimize supply chain operations. These innovations lead to greater efficiency, reduced errors, and improved decision-making, ultimately enhancing the quality of care provided to patients. By staying at the forefront of technological advancements, the global medical supply chain can continuously improve and support the overarching goals of healthcare stewardship (Blanchard, 2021, Khan et al., 2021)

4.7 The Global Medical-Industrial Complex

The global medical-industrial complex (GMIC) refers to the network of private and public organizations world-wide involved in the production and distribution of healthcare services delivery, durable medical equipment, pharmaceuticals, and advanced medical technologies (Rutkove et al., 2016). The GMIC encompasses hospitals, insurance companies, pharmaceutical manufacturers, durable medical supply distributors, medical device producers, academic and independent research institutions, and regulatory bodies. The GMIC is characterized by its significant integration of profit-driven motives with healthcare services delivery, leading to a complex interplay between market forces and the needs of patients around-the-globe. The GMIC structure has led to considerable innovation and advancements in medical technologies and treatments, often pushing the boundaries of what's possible in healthcare.

However, the GMIC has also been the subject of extensive criticism and debate. One major point of contention is the high cost of healthcare services and delivery in the United States, which is largely attributed to the profit-oriented nature of the industry (Salmon et al., 2021). Unlike many other developed nations, the U.S. lacks a universal healthcare system, resulting in significant disparities in access to care. The emphasis on profit can sometimes prioritize reactive care with expensive diagnostics, treatments, and procedures over more cost-effective, proactive care approaches. Additionally, the relationship between pharmaceutical companies and governmental regulatory bodies has raised concerns about the potential for conflicts of interest, particularly when it comes to drug pricing and the approval of new medications.

Moreover, the insurance industry plays a pivotal role in shaping the medical-industrial complex (Browner et al., 2019). The complex billing systems, negotiations over service prices, and the exclusion of certain treatments or medications from coverage plans are all influenced by the profit-driven motives of insurance providers. This often leaves patients with high out-of-pocket costs, even when they have insurance. The complexity and opacity of pricing within the system can also lead to unexpected and exorbitant medical bills, contributing to medical debt, which is a leading cause of bankruptcy in the U.S and many other countries around the world.

The GMIC also impacts the research and development sector (Blume, 2020). While the pursuit of new medical technologies and drugs is crucial for advancing healthcare, the heavy reliance on private funding can skew research priorities towards more profitable areas, sometimes at the expense of essential but less lucrative fields, such as vaccines or treatments for rare diseases. The influence of pharmaceutical companies and the prioritization of profits can also lead to issues such as the over-prescription of medications and the promotion of costly new treatments over established, effective, and cheaper alternatives.

The GMIC represents a vast and intricate system with both positive and negative aspects. While it has driven significant advancements in medical technology and treatments, it has also contributed to high healthcare costs, disparities in access, and ethical concerns regarding the prioritization of profit over patient welfare. Balancing the benefits of innovation with the need for equitable and affordable healthcare remains a key challenge for policymakers and stakeholders within this complex system.

The global medical supply chain is a fundamental component of the GMIC, providing the necessary infrastructure to deliver medical products and technologies from manufacturers to healthcare providers (Wieland, 2021). This chain encompasses various stages, including production, procurement, distribution, and delivery, all of which are integral to the functioning of the global healthcare delivery system. The efficiency and reliability of the medical supply chain ensure that hospitals, clinics, and other healthcare facilities have consistent access to essential supplies, such as medications, medical devices, and personal protective equipment. By ensuring the timely and adequate delivery of these supplies, the global medical supply chain supports the continuity of care and the effective treatment of patients.

Within the GMIC, the global medical supply chain also drives innovation and advancement (Salmon et al., 2021). Companies within the global medical supply chain invest heavily in research and development to create new medical products and improve existing ones. This innovation leads to the development of cutting-edge technologies and treatments that enhance patient outcomes and contribute to the overall progress of biomedical science. The competitive nature of the global medical supply chain fosters a continuous cycle of improvement and adaptation, ensuring that the healthcare delivery system benefits from the latest advancements and maintains high standards of care.

The economic impact of the global medical supply chain within the GMIC is significant (Salmon, 2022). It is a major driver of economic activity, supporting a vast network of manufacturers, suppliers, distributors, and logistics providers. This network creates jobs and stimulates economic growth, contributing to the financial stability of the global health sector and the broader global economy. Furthermore, the global medical supply chain's ability to negotiate pricing and manage costs directly influences the affordability of essential healthcare services. By optimizing procurement processes and reducing medical waste, the global medical supply chain helps to control the total cost of care, making diagnostics and treatments more accessible to patients and improving the overall efficiency of the healthcare delivery system.

Moreover, the global medical supply chain plays a critical role in global healthcare delivery systems by facilitating the distribution of medical products across borders (Dent, 2024). In the context of the GMIC, this global reach is essential for addressing health disparities and ensuring that medical innovations are available to populations worldwide. Through international partnerships and collaborations, the global medical supply chain supports global health initiatives, such as vaccination campaigns and disease eradication efforts. By enabling the global exchange of medical goods and knowledge, the global medical supply chain contributes to the improvement of health outcomes around-the-world.

Finally, the global medical supply chain is pivotal in ensuring emergency preparedness rapid response, and resilience within the GMIC (Nguyen, 2023). The ability to respond swiftly to public health emergencies, such as pandemics or natural disasters, relies heavily on a robust and agile global medical supply chain. Strategic stockpiling, diversified sourcing, and advanced logistics capabilities are essential for maintaining the flow of critical medical supplies during crises. By strengthening the resilience of the global medical supply chain, the GMIC can better withstand disruptions and continue to provide essential healthcare services, safeguarding the public's health and ensuring that global healthcare delivery systems remain functional under repeated episodes of stress.

4.8 References

1. Ahmad, F. B., Anderson, R. N. (2021) The leading causes of death in the US for 2020. *JAMA* 325(18), 1829-1830.
2. AlJaberi, O. A., Hussain, M., Drake, P. R. (2020) A framework for measuring sustainability in healthcare systems. *International Journal of Healthcare Management.*
3. Becker, C., Zumbrunn, S., Beck, K., et al. (2021) Interventions to improve communication at hospital discharge and rates of readmission: a systematic review and meta-analysis. *JAMA Network Open* 4(8): e2119346-e2119346.
4. Berry, L. L., Attai, D. J., Scammon, D. L., et al. (2022) When the aims and the ends of health care misalign. *Journal of Service Research* 25(1), 160-184.
5. Blanchard, D. (2021) *Supply chain management best practices.* John Wiley & Sons.
6. Blume, S. (2020) Medicine, technology and industry. *In* Medicine in the Twentieth Century (pp. 171-185). Taylor & Francis.
7. Browner, C. H. (2019) Moving beyond Neoliberal Models of Health Care. *Medical Anthropology* 38(5): 455-458.
8. Coombs, N. C., Meriwether, W. E., Caringi, J., et al. (2021) Barriers to healthcare access among US adults with mental health challenges: A population-based study. *SSM-PopulationHealth* 15:100847.
9. Debie, A., Khatri, R.B. & Assefa, Y. (2022). Successes and challenges of health systems governance towards universal health coverage and global health security: a narrative review and synthesis of the literature. *Health Res Policy Sys,* 20, 50.
10. Dent, M. (2024) The role of the medical–industrial complex in the National Health Services. *In* National Health Services of Western Europe (pp. 285-302). Routledge.
11. Fallahnezhad, M., Langarizadeh, M., & Vahabzadeh, A. (2024). Key performance indicators of hospital supply chain: A systematic review. *BMC Health Services Research.*

12. Fidler, D. P. (2017). *Emerging infectious diseases and global health security: The role of international law.* Global Health Governance, 10(1), 22-40.
13. Haque, M., McKimm, J., Sartelli, M., et al. (2020) Strategies to prevent healthcare-associated infections: a narrative overview. *Risk Management and Healthcare Policy* 13: 1765-1780.
14. Health at a Glance 2023: OECD Indicators. (2023). France: OECD Publishing.
15. Ketokivi, M., & Mahoney, J. T. (2020) Transaction cost economics as a theory of supply chain efficiency. *Production and Operations Management* 29(4): 1011-1031.
16. Khan, S. A. R., Godil, D. I., Jabbour, et al. (2021) Green data analytics, blockchain technology for sustainable development, and sustainable supply chain practices: evidence from small and medium enterprises. *Annals of Operations Research* pp 1-25.
17. Mekonnen, A. B., Redley, B., de Courten, B., et al. (2021) Potentially inappropriate prescribing and its associations with health-related and system-related outcomes in hospitalized older adults: a systematic review and meta-analysis. *British Journal of Clinical Pharmacology* 87(11): 4150-4172.
18. Mounier-Jack, S., Mayhew, S. H., & Mays, N. (2017). Integrated care: Learning between high-income, and low- and middle-income country health systems. *Health Policy and Planning, 32*(suppl_4), iv6–iv12.
19. Neglected Global Diseases Initiative (NGDI). (2023). *Global health access & equity issues in LMICs and HICs: Compare & contrast.*
20. Norton, W. E., Chambers, D. A. (2020) Unpacking the complexities of de-implementing inappropriate health interventions. *Implementation Science* 15(1), 2.
21. Nguyen, T. (2023) *Building health system resilience.* The Norwegian Defense Research Establishment.

22. Patel, B. S., Sambasivan, M. (2022) A systematic review of the literature on supply chain agility. *Management Research Review* 45(2): 236-260.
23. Rahman, M., Meyers, D. J., Wright, B. (2020) Unintended consequences of observation stay use may disproportionately burden Medicare beneficiaries in disadvantaged neighborhoods. *Mayo Clinic Proceedings* 95 (12): 2589-2592.
24. Rehman, O. U., Ali, Y. (2022) Enhancing healthcare supply chain resilience: decision-making in a fuzzy environment. *The International Journal of Logistics Management* 33(2): 520-546.
25. Rutkove, S. B., & Rutkove, S. B. (2016) Working with Industry. *Biomedical Research: An Insider's Guide*, 241-245.
26. Salmon, J. W. (2022) Organizing medical care for profit. *In* Issues in the political economy of health care (pp. 143-186). Routledge.
27. Salmon, J. W., Thompson, S. L., Salmon, J. W., et al. (2021) History of the Corporatization of American Medicine: The Market Paradigm Reigns. *The Corporatization of American Health Care: The Rise of Corporate Hegemony and the Loss of Professional Autonomy* pp 1-59.
28. UN General Assembly. (2015) *Transforming our world: the 2030 Agenda for Sustainable Development.*
29. Vahidi, S., Mirhashemi, S. H., Noorbakhsh, M., et al. (2020). Clinical errors: Implementing root cause analysis in an area health service. *International Journal of Healthcare Management* 13: 1-12.
30. Wieland, A. (2021) Dancing the supply chain: Toward transformative supply chain management. *Journal of Supply Chain Management*: 57(1), 58-73.
31. World Bank. (2022). *Digital health: Transforming healthcare delivery*. World Bank: Washington, D.C.
32. World Health Organization (WHO). (2022). *Universal health coverage*. World Health Organization.

33. World Health Organization (WHO). (2022). *Health system performance assessment: A framework for policy analysis.* World Health Organization.
34. World Health Organization (WHO). (2023). *Primary health care.* World Health Organization.
35. World Health Organization (WHO). (n.d.). *PHC measurement framework and indicators.* World Health Organization.
36. World Health organization (WHO). (2025). Governance and Universal Health Coverage. World Health Organization.
37. World Health Organization (WHO) & UNICEF. (2022). *Primary health care measurement framework and indicators: Monitoring health systems through a primary health care lens.* World Health Organization.
38. Yang, M., Fu, M., Zhang, Z. (2021) The adoption of digital technologies in supply chains: Drivers, process and impact. *Technological Forecasting and Social Change* 169: 120795.
39. Zhu, G., Chou, M. C., Tsai, C. W. (2020) Lessons learned from the COVID-19 pandemic exposing the shortcomings of current supply chain operations: A long-term prescriptive offering. *Sustainability* 12(14): 5858.
40. Ziat, A., Sefiani, N., Reklaoui, K., et al. (2020) A generic framework for hospital supply chain. *International Journal of Healthcare Management.*

5.0

A Brief Overview of the Global Public Health System.

The global public health system is a complex and interconnected network of institutions, frameworks, and actors dedicated to safeguarding and promoting the health of populations across the world. Rooted in principles of prevention, equity, and collaboration, this system operates at the intersection of diverse disciplines, including medicine, epidemiology, environmental science, policy, and community health. Its overarching goals encompass preventing disease, responding to health emergencies, and addressing the health-related social needs (HRSNs) to reduce health disparities and improve health literacy and health outcomes for all.

At the heart of the global public health system are international organizations, such as the World Health Organization (WHO), which coordinate efforts to tackle transnational health challenges. These institutions work in concert with national health authorities, non-governmental organizations, academic institutions, and the private sector to address a wide array of public health issues. From combating infectious diseases around-the-globe like HIV/AIDS, malaria, and tuberculosis to mitigating the impacts of chronic conditions and emerging global threats like human-induced climate change, the global public health system is built on the foundation of collective action, shared responsibility, health security, and social justice.

One of the defining features of the global public health system is its emphasis on disease surveillance, rapid response, and data-driven decision-making. Modern public health relies on robust monitoring mechanisms around-the-world to track disease patterns, detect outbreaks, and inform targeted interventions. Additionally, the global public health system is underpinned by a commitment to health equity, striving to ensure that vulnerable and underserved populations have access to essential healthcare services and resources. While it has achieved remarkable successes—such as the eradication of smallpox and progress in childhood immunization—ongoing challenges, including funding gaps, health disparities, and geopolitical tensions, underscore the need for continuous innovation and strengthened global solidarity.

5.1 The Essential Elements of the Global Public Health System.

The global public health system operates on a foundation of essential elements that enable it to address health challenges, promote health equity, and develop resilience on an international scale. These elements encompass a broad array of healthcare organizations, global healthcare delviery systems, academic and research institutions, and frameworks that collectively enhance the health, wellness, well-being, and resilience of populations worldwide. Through their integration, these components ensure the global public health system's ability to respond effectively to both enduring and emerging threats to the public's health world-wide.

1. *Disease Surveillance, Monitoring, and Response*: One of the cornerstones of the global public health system is robust disease surveillance. Monitoring mechanisms track disease prevalence,

identify outbreaks, and assess health trends in real time. These systems facilitate a rapid response, when indicated, with timely interventions and evidence-based healthy public policymaking. The World Health Organization (WHO) plays a central role in this domain, collaborating with regional health networks to share data and coordinate responses (WHO, 2023).

2. *Health Equity and Determinants of Health*: A commitment to health equity underscores the global public health framework. Addressing disparities, the global public health system targets the health-related social needs (HRSNs) and other determinants of health—such as housing, food, transportation, education, income, and access to clean air, water, and soil—to improve outcomes for underserved populations (Marmot & Allen, 2020). Interventions often focus on creating access to essential healthcare services and fostering inclusive healthy public policies.
3. *Emergency Preparedness and Response*: Preparedness for pandemics, natural disasters, and other crises is a defining feature of the global public health system. This includes establishing international response protocols, such as those outlined in the International Health Regulations (IHR) (2005), to contain infectious disease threats and mitigate their impact (WHO, 2005). Agencies like the Centers for Disease Control and Prevention (CDC) also contribute through technical support and expertise world-wide (CDC, 2022).
4. *Health Prevention, Protection, and Promotion*: Preventive measures are foundational to the goals of the global public health system, serving as a proactive approach to mitigating health risks and promoting the wellness and well-being of populations. By prioritizing vaccination programs, health education initiatives, and community outreach, the global public health system strives to reduce the incidence of disease and foster healthier lifestyles worldwide.

   Vaccination campaigns remain among the most powerful tools in health protection, preventing the spread of infectious diseases that have historically posed significant threats to humanity. Landmark achievements such as the eradication of smallpox and the near-elimination of poliomyelitis highlight the profound impact of immunization programs on global health outcomes. These successes underscore the importance of sustaining vaccination efforts to address emerging infectious disease threats, including outbreaks of measles, influenza, and other vaccine-preventable conditions. Organizations such as Gavi, the Vaccine Alliance (2023), play a pivotal role in ensuring equitable access to vaccines, particularly for underserved communities.

Equally vital are health education initiatives, which engage and empower individuals and communities to make informed choices about their health, wellness and well-being. Global public health system campaigns focus on raising awareness about lifestyle and behavioral factors that influence health, such as nutrition and diet, physical activity, sleep, stress, and illicit substance use. For example, efforts to combat tobacco consumption through education and healthy public policy interventions have significantly reduced smoking prevalence and associated health risks. Similarly, programs promoting healthy eating habits and exercise contribute to preventing chronic conditions such as obesity, diabetes, and cardiovascular disease. Beyond individual behavior, preventive strategies extend to addressing HRSNs and environmental determinants of health. This includes advocating for clean water and sanitation, combating air pollution, and improving housing conditions. In tackling these systemic factors, global public health initiatives create an enabling environment where prevention can thrive. Moreover, integrating health promotion into broader healthy public policies—such as urban planning and climate action—amplifies the reach and impact of these measures. In summation, health protection, prevention, and promotion form the backbone of the global public health system's efforts to safeguard populations from illness and enhance quality of life. By combining vaccination campaigns, health education, and systemic interventions, the global public health system not only reduces disease prevalence but also fosters resilience and equity in the face of evolving health challenges.

5. International Collaboration and Governance: Collaboration among nations, international organizations, and non-governmental entities underpins the global public health system's effectiveness. Frameworks such as the United Nations' Sustainable Development Goals (SDGs) highlight the importance of global solidarity in achieving universal health coverage (UHC) and advancing public health objectives (UN, 2015).

In conclusion, the global public health system's essential elements form an intricate web of interconnected functions and initiatives. By focusing on disease surveillance and monitoring, equity, emergency preparedness and rapid response, health protection, prevention, and promotion, governess and collaboration, the global public health system seeks to address complex international health challenges in an increasingly interdependent world. While considerable progress has been made, continued investment and innovation are imperative to ensure its resilience and adaptability to future existential threats to all living beings on the planet.

5.2 Innovations in the Global Public Health System.

Public health innovations have significantly transformed the landscape of health and wellness, improved outcomes and enhanced the quality of life on a global scale (Rilkoff et al., 2024). These innovations span various domains, from advanced medical technology and data analytics to healthy public policy and community-based interventions, each contributing to the advancement of global public health initiatives.

One of the most significant global public health innovations is the development and widespread implementation of vaccines (Cable et al., 2020). Vaccination programs have been instrumental in controlling and eradicating infectious diseases such as smallpox, polio, and measles. The rapid development and distribution of COVID-19 vaccines is a recent example, demonstrating the power of scientific advancements and global collaboration in addressing public health crises. These efforts not only save millions of lives but also prevent the spread of diseases, contributing to the overall health of populations. More critically, shifting the focus from reactive care—addressing acute illnesses, injuries, and the management of chronic diseases—to proactive, predictive, and personalized care alleviates the burden placed on an already overstretched global healthcare delivery system. By emphasizing early intervention and wellness strategies, this approach not only reduces the prevalence of preventable conditions but also enhances overall system efficiency and sustainability.

Telehealth and digital health technologies represent another major innovation in global public health systems (Gunasekeran et al., 2021). Telehealth allows for remote consultations and monitoring, expanding access to essential healthcare services, especially in underserved and rural areas. Mobile health applications and wearable devices enable individuals to track their health metrics, manage chronic conditions, and receive timely health information. These advanced medical technologies empower patients to take control of their health and facilitate more efficient and effective healthcare delivery.

Data analytics and big data have revolutionized the global public health system by enhancing the ability to predict, monitor, and rapidly respond to modern-day global existential health threats. Advanced data analytics tools allow international public health professionals to identify patterns and trends in health data, enabling proactive measures and targeted interventions. For instance, predictive analytics can forecast emerging infectious disease outbreaks, helping global public health authorities to implement preventive measures swiftly. Additionally, big data can inform healthy public policy decisions and resource allocation, ensuring that public health initiatives are value-driven, evidence-based and impactful (Sarker et al., 2021).

Community-based interventions and community-based participatory approaches are also key innovations in global public health delivery (Nickel et al., 2020; Faber et al., 2021). Programs that engage communities in the planning and implementation of health initiatives are more likely to be successful and sustainable. Examples include community health worker programs, which leverage local community knowledge and trust to deliver health education and services, and participatory research projects that involve community members in identifying health priorities and developing solutions (Kane et al., 2021). These approaches foster a sense of ownership and empowerment, leading to more effective and culturally appropriate global public health interventions.

Finally, advancements in genomics and P4 (predictive, personalized, preventive, and participatory) medicine are paving the way for more tailored global public health strategies (Yurkovich et al., 2024). Understanding the genetic factors that influence health and disease can lead to more precise and personalized prevention and treatment plans. For instance, genetic screening programs can pinpoint individuals at elevated risk for severe and debilitating diseases. This early identification enables timely interventions, the development of tailored care plans, and, in some cases, the application of advanced treatments such as cell and gene therapies. These measures collectively enhance health outcomes, extend health spans, and contribute to significant reductions in healthcare costs.

In summary, global public health innovations in vaccines, digital health, data analytics, community-based interventions, and genomics are driving significant improvements in global health outcomes. These advancements enable more efficient, effective, and equitable global public health strategies, ultimately contributing to the health, wellness, and well-being of the global population-at-large.

5.3 The Vital Role of a Resilient Global Public Health System Centered on Healthcare Stewardship.

The global public health system plays a critical role in healthcare stewardship by serving as a central hub for coordinating efforts across various healthcare providers and community organizations (Aragon et al., 2015). They help develop and implement public health policies and healthy public policies and programs that ensure the efficient use of limited healthcare resources while promoting the health, wellness, well-being, and resilience of the population. Through comprehensive community health needs assessments and the collection of health outcome data, public health agencies at local, state, national, regional, and international levels can pinpoint areas with the greatest needs. This enables the strategic allocation of resources and the development of targeted interventions to address specific challenges. Such data-driven planning enhances efficiency, minimizes resource wastage, and ensures that efforts are concentrated on initiatives with the highest potential for impact.

In addition, global public health agencies are instrumental in promoting proactive care and health education, which are key components of healthcare stewardship. The agencies' campaigns and initiatives to raise awareness about the importance of vaccination, regular health screenings, healthy lifestyles, and disease prevention. By educating the public and encouraging proactive health management, global public health agencies help to reduce the incidence of chronic conditions that can place a significant burden on global healthcare delivery systems. This proactive approach not only improves individual health outcomes but also decreases the overall demand for healthcare services, thus optimizing the use of available healthcare resources (Caron et al., 2023).

Global public health agencies also play a vital role in managing and responding to public health emergencies, such as infectious disease outbreaks and natural disasters. Their ability to quickly mobilize resources, disseminate information, and coordinate with various stakeholders is crucial in mitigating the impact of such events. Effective emergency preparedness, monitoring, and response plans developed at all levels of the global public health system ensure that healthcare resources are used efficiently and that the community receives timely and appropriate healthcare services. This level of preparedness is a fundamental principle of healthcare stewardship, as it helps to prevent the overwhelming of global healthcare delivery systems during crises (Rose et al., 2017).

Furthermore, global public health agencies foster collaboration and partnerships among different sectors, including healthcare providers, government healthcare agencies, non-profit organizations, and the private sector. By bringing together diverse stakeholders, they create a more integrated and cohesive global healthcare delivery system that can share resources, knowledge, and best practices. This collaborative approach enhances the overall effectiveness and efficiency of healthcare services delivery, ensuring that resources are used wisely and that patients receive value-driven, cost-effective, and high-quality care. Public health agencies at all levels of the global system act as catalysts for innovation and continuous quality improvement, continually seeking ways to enhance healthcare stewardship through education, practice, research, policy development, and community engagement.

5.4 The Essentials of Community-based Organizations.

Community-based organizations (CBOs) play an essential role in addressing local needs and enhancing the quality of life within communities (Wilson et al., 2012). These organizations are typically non-profit entities that operate at the grassroots level, providing a wide range of clinical and non-clinical services and support tailored to the specific needs of their communities. By leveraging local knowledge and resources, CBOs are uniquely positioned to respond effectively to various social, economic, and health challenges.

One of the primary functions of CBOs is to offer direct services and support to community members (Taylor et al., 2021). This can include providing food assistance, housing support, job training, healthcare services, and educational programs. For instance, food banks and meal programs run by CBOs ensure that individuals and families facing food insecurity have access to nutritious meals. Similarly, CBOs offering housing assistance help prevent homelessness by providing shelter, rental assistance, and supportive housing services. These direct services are crucial in meeting the immediate needs of vulnerable populations and promoting overall community well-being.

Another critical function of CBOs is advocacy and community organizing (Truong et al., 2023). CBOs often serve as advocates for their communities, working to influence healthy public policy and bring about systemic changes that address the root causes of social and economic inequalities. By mobilizing community members and building coalitions, CBOs can amplify the voices of marginalized groups and push for public policy reforms that promote social justice and equity. For example, CBOs may advocate for affordable housing policies, improved access to healthcare services, or educational reforms that benefit underserved communities.

CBOs also play a vital role in fostering social cohesion and community engagement (Habermann et al., 2014). They provide spaces for community members to come together, share experiences, and collaborate on initiatives that improve their neighborhoods. Through community meetings, cultural events, and volunteer opportunities, CBOs help build a sense of belonging and collective responsibility among residents. This social capital is essential for creating resilient communities that can effectively address challenges and leverage opportunities for growth and development.

Furthermore, CBOs often serve as intermediaries between communities and larger institutions, such as government agencies, healthcare organizations, and philanthropic institutions (Franks et al., 2017). By acting as a bridge, CBOs ensure that community needs and perspectives are considered in decision-making processes. They also help facilitate access to resources and essential healthcare services provided by these institutions, ensuring that they reach the individuals who need them the most. For instance, CBOs might partner with local healthcare providers to offer community health screenings or work with local governments to implement neighborhood improvement projects.

In summary, community-based organizations are indispensable in establishing a sense of belonging by addressing the diverse needs of their communities through direct services, advocacy, community engagement, and acting as intermediaries with larger institutions. Their grassroots approach and deep understanding of local contexts enable them to effectively respond to challenges, promote social cohesion and belonging, and drive sustainable positive change.

5.5 Engaging Community-Based Organizations in Healthcare Stewardship.

Engaging community-based organizations (CBOs) in healthcare stewardship is a powerful strategy to enhance the effectiveness and reach of healthcare initiatives. CBOs have deep roots within their communities and possess a thorough understanding of local needs, cultures, and challenges. By collaborating with these organizations, healthcare providers can leverage this local knowledge to design and implement more effective health programs and interventions. CBOs can serve as trusted intermediaries, helping to bridge the gap between healthcare systems and the populations they serve, ensuring that healthcare services are accessible, acceptable, and appropriate for diverse community members (Franks et al., 2017).

One of the key benefits of involving CBOs in healthcare stewardship is the ability to improve health literacy and empower individuals to take an active role in managing their health (Palumbo et al., 2018; Sorenson et al., 2021). CBOs often have established communication channels, trustworthiness, and truthfulness within the community, making them ideal partners for disseminating information about proactive health, healthy lifestyles and behaviors, and available essential healthcare services. This value-driven grassroots approach can lead to better health outcomes and reduced healthcare costs by preventing illnesses and managing chronic conditions more effectively.

CBOs also play a critical role in addressing all determinants of health, which are key factors in healthcare stewardship (Taylor et al., 2021). Issues such as housing, education, employment, and food security significantly impact health outcomes. CBOs are often involved in these areas and can provide comprehensive support that addresses both health and health-related social needs. By partnering with CBOs, healthcare providers can adopt a holistic, integrated, and comprehensive approach to people-centered care, mixing social services with advanced primary health care to address the root causes of health issues. This collaborative approach can lead to more environmentally sustainable health improvements and a more efficient use of healthcare resources.

Furthermore, engaging CBOs can enhance the cultural competence of healthcare delivery (Garrido et al., 2019). CBOs often have staff and volunteers who are part of the communities they serve, bringing valuable cultural insights and language skills to healthcare interactions. This cultural competency is crucial for building trust, improving patient-provider communication, and ensuring that healthcare services are culturally appropriate. By incorporating the perspectives and expertise of CBOs, healthcare providers can develop and deliver healthcare services that are more responsive to the cultural needs of diverse populations, leading to better patient satisfaction and adherence to treatment plans.

Involving CBOs in healthcare stewardship also facilitates community-driven health initiatives, empowering communities to take ownership of their health outcomes (Schensul et al., 2014). CBOs can help organize and support community health boards, advisory councils, and participatory research projects, giving community members a voice in the planning, implementation, and evaluation of community-based health programs. This participatory approach ensures that health initiatives are aligned with community priorities and are more likely to be environmentally sustainable in the long term. It also fosters a sense of ownership and accountability among community members, encouraging active participation in proactive health services efforts.

Lastly, partnerships with CBOs can strengthen the overall resilience of healthcare delivery systems (Koch et al., 2017). During public health emergencies, such as natural disasters or infectious disease outbreaks, CBOs can provide essential support by mobilizing community resources, disseminating valid and reliable critical health and non-health related information, and coordinating local response efforts. Their established presence and trust within the community enable them to act swiftly and effectively, complementing the efforts of the medical community and local public health districts. By integrating CBOs into emergency preparedness and response plans, healthcare delivery systems, public health agencies, and emergency management systems can enhance their capacity to respond to crises and ensure continuity of care for marginalized individuals and vulnerable populations.

In summary, engaging CBOs in healthcare stewardship brings numerous benefits including improved health literacy, holistic, integrated, and comprehensive healthcare services addressing all the upstream determinants of health, enhanced cultural competence, community-driven health initiatives, and strengthened multi-system resilience. These partnerships create a more inclusive, effective, and sustainable healthcare delviery system that is better equipped to meet the needs of diverse communities and address the complex challenges of modern healthcare.

5.6 Collaborative Pathways: Public Health and Community-Based Organizations in Healthcare Stewardship.

The relationship between public health agencies and community-based organizations (CBOs) is dynamic and critical for addressing the health needs of populations effectively (Acosta et al., 2018). Public health agencies, including local, state, and national health departments, are central to advancing and safeguarding community health through initiatives such as healthy public policy formulation, disease surveillance and response systems, health education campaigns, and proactive measures to prevent diseases and promote health, wellness, well-being, and resilience. Meanwhile, community-based organizations (CBOs) function at the grassroots level, offering a closer connection to the populations they serve. With their deep integration within communities, CBOs often focus on delivering specialized health services, advocating for underserved groups, and providing social support tailored to specific needs or addressing distinct health challenges. Together, these entities create a complementary framework for promoting equitable and effective healthcare stewardship.

One of the key dynamics in this relationship is collaboration and partnership. Public health agencies often collaborate with CBOs to leverage their expertise, community trust, and cultural competence in implementing health interventions (Agonafer et al., 2021). CBOs, in turn, benefit from public health resources, technical support, and access to data and evidence-based clinical practices that enhance their effectiveness. Effective collaboration fosters a shared understanding of community needs, strengthens the capacity to address health disparities, and maximizes the impact of interventions by combining resources and expertise.

Another dynamic is the role of CBOs in advocating for community health needs and influencing healthy public policies (Allen et al., 2021). CBOs are uniquely positioned to advocate for policies that address the determinants of health, promote health equity, and address the specific needs of marginalized or underserved populations (Winton et al., 2018). They serve as a bridge between communities and public health authorities, advocating for community priorities and ensuring that public health initiatives are culturally responsive and inclusive. This advocacy role is crucial in shaping healthy public policies that support community well-being and foster sustainable improvements in population health outcomes.

Challenges also exist in this relationship, including resource constraints, communication barriers, and differing organizational priorities. Public health agencies may face limitations in funding and capacity, impacting their ability to support and sustain partnerships with CBOs effectively. Communication challenges, such as differing terminology or bureaucratic processes, can hinder effective collaboration and coordination of efforts. Moreover, aligning organizational priorities and goals between public health agencies and diverse CBOs with varying missions and objectives requires ongoing dialogue, mutual respect, and commitment to shared goals (Yasmin et al., 2022).

Overall, navigating the relationship between public health and CBOs involves recognizing and leveraging each other's strengths, fostering trust, truthfulness, and collaboration, and addressing challenges through effective communication and strategic partnerships. By working together synergistically, public health agencies and CBOs can enhance their collective impact to produce positive health, optimize wellness, enhance the well-being of communities, and develop sustainable resilience particularly those communities facing health disparities and social inequities.

5.7 References.

1. Acosta, J. D., Burgette, L., Chandra, et al. (2018) How community and public health partnerships contribute to disaster recovery and resilience. *Disaster Medicine and Public Health Preparedness* 12(5): 635-643.
2. Agonafer, E. P., Carson, S. L., Nunez, V., et al. (2021) Community-based organizations' perspectives on improving health and social service integration. *BMC Public Health* 21: 1-12.
3. Allen, E. H., Haley, J. M., Aarons, J., et al. (2021) *Leveraging community expertise to advance health equity.* Washington, DC: Urban Institute.

4. Aragón, T. J., Garcia, B. A., et al. (2015) Designing a learning health organization for collective impact. *JPHMP* 21(Suppl 1): S24–S33.
5. Cable, J., Srikantiah, P., Crowe Jr, et al. (2020) Vaccine innovations for emerging infectious diseases—a symposium report. *Annals of the New York Academy of Sciences* 1462(1): 14-26.
6. Caron, R. M., Noel, K., Reed, R. N., et al. (2023) Health Promotion, Health Protection, and Disease Prevention: Challenges and Opportunities in a Dynamic Landscape. *AJPM* 3(1): 100167.
7. Centers for Disease Control and Prevention (CDC). (2022). International Preparedness and Response. Centers for Disease Control and Prevention.
8. DeSalvo, K. B., Kadakia, K. T. (2021). Public Health 3.0 after COVID-19—reboot or upgrade? *American Journal of Public Health* 111(S3): S179-S181.
9. Faber, J. S., Al-Dhahir, I., Reijnders, T., et al. (2021). Attitudes toward health, healthcare, and eHealth of people with a low socioeconomic status: a community-based participatory approach. *Frontiers in Digital Health* 3: 690182.
10. Franks, R. P., Bory, C. T. (2017) Strategies for developing intermediary organizations: Considerations for practice. *Families in Society* 98(1): 27-34.
11. Garrido, R., Garcia-Ramirez, M., Balcazar, F. E. (2019) Moving towards community cultural competence. *International Journal of Intercultural Relations* 73: 89-101.
12. Gavi, The Vaccine Alliance (GAVI). (2023). Preventive Health.
13. Goldsteen, R. L., Goldsteen, K., Dwelle, T. (2024). Introduction to public health: promises and practices. 2nd Edition. New York, NY: Springer Publishing.
14. Gunasekeran, D. V., Tseng, R. M. W. W., Tham, et al. (2021) Applications of digital health for public health responses to COVID-19: a systematic scoping review of artificial intelligence, telehealth and related technologies. *NPJ Digital Medicine* 4(1): 40.

15. Habermann, H., Mackie, C. D., Prewitt, K. (Eds.). (2014) *Civic engagement and social cohesion: Measuring dimensions of social capital to inform policy* Washington, DC: National Academies Press.
16. Kane, S., Radkar, A., Gadgil, M., et al. (2021) Community health workers as influential health system actors and not" just another pair of hands". *International Journal of Health Policy and Management* 10(8): 465.
17. Koch, H., Franco, Z. E., O'Sullivan, T., et al. (2017) Community views of the federal emergency management agency's "whole community" strategy in a complex US City: Re-envisioning societal resilience. *Technological Forecasting and Social Change* 121: 31-38.
18. Marmot, M., & Allen, J. (2020). Social Determinants of Health: Equity and the Global Health System. *The Lancet.*
19. Nickel, S., von dem Knesebeck, O. (2020) Effectiveness of community-based health promotion interventions in urban areas: a systematic review. *Journal of Community Health* 45(2): 419-434.
20. Palumbo, R., Annarumma, C. (2018) Empowering organizations to empower patients: An organizational health literacy approach. *International Journal of Healthcare Management* 11(2): 133-142.
21. Rilkoff, H., Struck, S., Ziegler, C., et al. (2024) Innovations in public health surveillance: An overview of novel use of data and analytic methods. *Canada Communicable Disease Report* 50(3-4): 93–101.
22. Rose, D. A., Murthy, S., Brooks, J., et al. (2017) The Evolution of Public Health Emergency Management as a Field of Practice. *American Journal of Public Health* 107(S2): S126–S133. https://doi.org/10.2105/AJPH.2017.303947
23. Rosen, G. (2015). *A history of public health.* Jhu Press.
24. Sarker, I. H. (2021) Data science and analytics: an overview from data-driven smart computing, decision-making and applications perspective. *SN Computer Science* 2(5): 377.
25. Schensul, S. L., Schensul, J. J., Singer, M., et al. (2014) Participatory methods and community-based collaborations. *Handbook of Methods in Cultural Anthropology* 2:185-212.

26. Sorensen, K., Levin-Zamir, D., Duong, T. V., et al. (2021) Building health literacy system capacity: a framework for health literate systems. *Health Promotion International* 36(Suppl): i13-i23.
27. Taylor, L. A., Byhoff, E. (2021) Money moves the mare: the response of community-based organizations to health care's embrace of social determinants. *The Milbank Quarterly* 99(1): 171-208.
28. Truong, J., Sandhu, P., Sheng, et al. (2023) Advocacy in community-based service learning: perspectives of community partner organizations. *Canadian Medical Education Journal* 14(1): 90-94.
29. United Nations (UN). (2015). Sustainable Development Goals. United Nations: New York, NY.
30. Wilson, M. G., Lavis, J. N., Guta, A. (2012) Community-based organizations in the health sector: a scoping review. *Health Research Policy and Systems* 10: 1-9.
31. Winton, S., Evans, M. P. (2018) Consulting, mediating, conducting, and supporting: How community-based organizations engage with research to influence policy. Pp 4-25. *In* Engaging Families, Educators, and Communities as Educational Advocates: Routledge.
32. World Health Organization (WHO). (2005). International Health Regulations. World Health Organization.
33. World Health Organization (WHO). (2023). Global Surveillance and Monitoring. World Health Organization.
34. Yasmin, S., Haque, R., Kadambaya, K., et al. (2022) Exploring how public health partnerships with community-based organizations (CBOs) can be leveraged for health promotion and community health. *INQUIRY: The Journal of Health Care Organization, Provision, and Financing* 59: 00469580221139372.
35. Yurkovich, J. T., Evans, S. J., Rappaport, N., et al. (2024) The transition from genomics to phenomics in personalized population health. *Nature Reviews Genetics* 25(4): 286-302.

6.0

The Global Medical Commons.

6.1 The *Tragedy of the Commons.*

The "*Commons*" refer to shared resources that are accessible to all members of a community or society, often without private ownership. These resources can include natural elements like air, water, forests, and fisheries, as well as intangible assets such as knowledge, public spaces, and cultural heritage. The concept of the commons emphasizes collective stewardship and responsibility, ensuring that these resources are preserved and managed sustainably for present and future generations. In practice, the governance of commons often involves cooperative decision-making and strategies to prevent overuse or degradation, a challenge widely known as the "*tragedy of the commons.*"

In his landmark 1968 article, *The Tragedy of the Commons*, published in *Science*, Garrett Hardin examined a category of human challenges that inherently defy technical solutions. In the article, he defined a technical solution as "*...one that requires a change only in the techniques of the natural sciences, demanding little or nothing in the way of change in human values or ideas of morality.*" Hardin was discussing the issues related to population growth. He summarized that even with an infinite amount of energy available to sustain an ever-expanding population, individual and community interests compete for these energy resources. This results in behavioral issues, both positive and negative, that cannot sustain the population forever. Hardin developed a hypothetical case study focused around a conundrum that remains appropriate for discussing health care stewardship in the twenty-first century.

Hardin begins to explain the reasons behind the initial development of the tragedy of the commons. Imagine a pasture accessible to everyone in a clearly defined geographic area. It is expected that each herdsman will try to keep as many cattle as possible on the pasture, known as the "*commons*." This arrangement may work reasonably well for centuries because tribal wars, poaching, and disease keep the numbers of both people and livestock well below the land's carrying capacity. Eventually, however, the day of reckoning arrives—the day when the long-desired goal of social stability becomes a reality. At this point, the inherent logic of the commons inevitably leads to tragedy.

Assuming the herdsmen are acting rationally, each herdsman seeks to maximize his gain. Consciously or unconsciously, they ask themselves, "What is the benefit to me of adding one more animal to my herd on the commons?"

This benefit has both a positive and a negative component. The positive component comes from the incremental addition of one animal to the commons. Since the herdsman receives all the proceeds from the sale of the additional animal, the positive benefit is nearly +1. The negative component results from the additional overgrazing caused by adding one more animal to the commons. Since the effects of overgrazing are shared by all herdsmen, the negative impact on any individual herdsman is only a fraction of -1. When combining these partial utilities, the rational herdsman concludes that the only sensible course of action is to add another animal to his herd growing in number on the commons. And then another; and yet another. However, this conclusion is reached by every rational herdsman sharing the commons. This is where the tragedy lies.

Each man is locked into a system that compels him to increase his herd without limit—in a world that is limited at best. Hardin (1968) concludes the "ruin" is the destination toward which all men rush, each pursuing his own best interest in a society that believes in the freedom of the commons. Freedom in a commons brings ruin to all.

In 1975, Dr. Howard H. Hiatt, an American biomedical scientist and one of the discoverers of mRNA, advanced the conversation about the tragedy of the commons by connecting it to the management of finite healthcare resources. He highlighted the "*tragedy of the healthcare commons*", emphasizing the challenges in balancing limited resources within the American healthcare delivery system. Hiatt suggested that the total resources available for medical care are analogous to the grazing area in Hardin's commons, while the medical practices utilizing those resources are similar to Hardin's grazing animals.

Dr. Hiatt observed that no one would dispute the existence of a limit to the resources any society can allocate to medical care, and by 1975, it was widely acknowledged that society was rapidly nearing this limit. He also pointed out that a significant portion of the increasing demands on healthcare resources was aimed at addressing the inadequacies of medical care for large segments of the population. In discussing the challenges of overpopulation and finite energy resources, Hiatt posed the critical question: "*How can we prevent the additional strain on the medical commons from leading to societal ruin?*"

6.2 Protecting the Global Medical Commons.

Competing interests among stakeholders in the global healthcare delivery system place immense strain on the global medical commons, pushing it to its limits. The lack of access to real-time clinical data from the point of care hampers efforts to prioritize and shape effective public policy. Furthermore, the high prevalence of chronic conditions heavily influences the allocation of healthcare resources, concentrating efforts on managing individual patients. This singular focus often detracts from addressing the broader health needs of the population as a whole, creating challenges in achieving a balanced and equitable approach to healthcare services delivery.

As it stands, the marginal benefits gained for individual patients can jeopardize the overall health, wellness, and well-being of the entire population. Unrestricted access to the medical commons, in both HICs and LMICs, has led to the imprudent use of scarce healthcare resources, which serves neither individuals nor society effectively. Evidence-based medicine and clinical practice guidelines, derived from randomized controlled trials, have prompted organized medicine to move away from practices that fall into this category.

The health, wellness, well-being, and resilience of the public-at-large have long been central to the mission of global healthcare delivery systems, encompassing both individual patient care in local clinics and the broader management of population health on a global scale. However, these efforts are increasingly shaped by competing paradigms: salutogenesis, which emphasizes the origins of health and the factors that foster well-being, and pathogenesis, which focuses on diagnosing and treating disease and injury.

The tension between these approaches has, over time, contributed to the overuse of the medical commons—shared healthcare resources that are critical to addressing the needs of both individuals and populations. As a result, the practice of modern medicine has become more expansive and complex, requiring integrated strategies that balance proactive health care with the demands of reactive care. This dynamic is reflected in healthcare systems worldwide, from local to global contexts, as they strive to navigate these competing priorities while managing resource limitations and rising demands.

Dr. Hiatt, recognizing the potential consequences of overusing the medical commons, posed the question: "Who will protect the medical commons?" At the turn of the twentieth century, Hiatt observed that the medical commons featured relatively few expressive clinical practices, lacked well-defined limits on the use of limited healthcare resources, and relied on conscientious practitioners and the cottage-like industry of medicine at that time to draw only what was necessary and essential for patient care.

By the twenty-first century, the global healthcare delivery system had undergone several transformations, culminating in today's chronic condition model of care. In addition, global healthcare services delivery has become increasingly subordinate to and influenced by the global medical-industrial complex (GMIC). A multitude of complex factors have stretched the boundaries of the medical commons, introducing significant existential challenges to public health on both national and global scales. These pressures threaten the capacity of healthcare delivery systems globally to address the diverse and evolving needs of populations effectively, underscoring the urgency for sustainable solutions and collective stewardship.

These threats encompass human-induced climate change accompanied by escalating greenhouse gas emissions, a surge in natural disasters, global pandemics such as COVID-19, and the persistent impacts of institutional bias and systemic racism. They also include deepening economic inequities, widening health disparities, an aging global population, the resurgence of infectious diseases, and a dramatic increase in chronic health conditions like diabetes, cardiovascular disease, and neurodegenerative disorders. Together, these challenges present profound and multifaceted risks to the health and resilience of populations worldwide.

Keeping the focus on protecting the medical commons, many questions develop to help avert a major tragedy including:

1. How should priorities be set?
2. Who should set these priorities?
3. How much should be allocated for health care resources?
4. Of the total medical commons, how much should be allocated for:
    a) Medical care?
    b) Basic science?

   c) Medical education?
   d) Public health?
   e) Medical research?
5. If a disease management program receives a limited sum of funding, how should that money be optimally used?
6. Who should qualify for treatment of their chronic end-stage diseases (e.g., end-stage renal disease and dialysis)?
7. Of all the stakeholders in the global healthcare delivery system, who should make health care resource allocation decisions? Healthcare providers? The government? The medical-industrial complex?
8. On what basis should health care resource allocation decisions be made?

Hiatt (1975) identified critical issues pertaining to the medical commons that remain highly relevant to the current landscape of global healthcare services delivery. First, he emphasized that health prevention and health promotion represent the most promising long-term strategies for safeguarding the limited resources of the medical commons. By prioritizing proactive measures, societies can reduce the strain on shared healthcare resources and enhance overall health outcomes.

Second, Hiatt urged clinicians to actively educate and encourage the public to deepen their understanding of individual health, wellness, well-being, and resilience. He believed that if individuals became more aware of the limitations and uncertainties inherent in healthcare, they would make more informed decisions about resource utilization. This awareness would highlight the importance of maintaining good health as a means of reducing reliance on healthcare systems.

Lastly, Hiatt underscored the necessity of flexible decision-making processes regarding the management of the medical commons. He advocated for the continuous evaluation and adaptation of healthcare strategies in response to evolving insights into what proves effective at various organizational levels—from sub-molecular and cellular to ecosystem-wide scales. Using the medical commons responsibly demands ongoing quality and process improvements, ensuring that healthcare delivery systems are both efficient and respectful of shared resources.

6.3 Managing Limited Medical Resources in the 21st Century U.S. Healthcare Delivery System

Unfortunately, the issues related to the medical commons that Hiatt so eloquently identified in 1975 remain unresolved today. Market-based attempts to control health care resource utilization by instituting the principles and practices of managed care and utilization management did nothing to limit the irresponsible use of health care resources by all the healthcare stakeholders. In fact, it is common knowledge that the U.S. spends more per capita on health care than any other developed or developing nation without achieving the highest quality, safety, and outcomes for the patients served (OECD, 2023).

Managing health care resources today is really a question of cost-efficiency vs. quality of and safety of care. The stakeholders of any healthcare delivery system around the globe are interested in efficiency, but it is the clinicians who struggle with their ongoing role in managing the appropriate use of health care resources. This struggle remains deeply rooted in medical ethics (Cassel and Brennan 2007).

The traditional core value of medicine is the primacy of the patient's well-being over the self-interest of the clinician and implicitly over other social concerns as well (Cassel et al., 2007). This is highly desirable when very ill patients seek medical attention from an expert clinician. There is a fundamental inequality in the fund of knowledge and power between the clinician and patient. In dire situations, patients acquiesce to the clinical expertise of the clinician whom they believe has their best interests at heart.

Should the clinician be concerned about, as well as committed to, managing the medical resources (i.e., medical commons) for his or her patients? The answer is yes. Today, it is impossible to avoid the fact that clinicians live and work in a medical commons and bear responsibility for making resource decisions that directly affect its utilization (Cassel et al., 2007). However, according to Starr (1982), the allopathic medical community worked aggressively to assert its control over the U.S. healthcare delivery system by partnering with the U.S. medical-industrial complex and further widening the affiliation between itself and public health sectors, while championing fee-for-service payment.

A basic appreciation for the medical commons escapes many clinicians today. Healthcare resource utilization, both directly and indirectly, benefit patients in need of health care goods and services. A good example of this phenomenon is the implementation of the managed care model introduced into the U.S. healthcare delivery system in the 1980s fostered this attitude (Daniels 1986). In fact, it became widely understood within the allopathic medical community that any savings from market-based activities would likely be used either to reduce the deficit in government-funded programs (e.g., Medicare and Medicaid) or to enhance the profit margins of the medical-industrial complex (e.g., PHARMA). Clinicians faced ethical and moral challenges with the rationing of the medical commons at the point of care, known as bedside rationing, because it compromised the sacred clinician-patient relationship and associated clinicians with cost-control measures aimed at benefiting payers' business interests rather than improving patient care.

To date, clinicians have largely relinquished responsibility for the medical commons in the United States to two major forces: market-based dynamics and government regulation. Can either of these forces effectively manage the medical commons without clinician involvement? The delivery of traditional medical care and the essential public health services over the past 50 years suggests strongly the answer would be no.

Government regulation, particularly with regard to the Medicare system at the federal level, has significantly contributed to the irresponsible use of limited healthcare resources within the U.S. healthcare delivery system. The oversight of physician service costs, based on the now-defunct (replaced in 2015 by MACRA) annual Medicare Sustainable Growth Rate (SGR) calculations, illustrated this issue year over year. Each winter, lobbyists from major professional medical organizations worked to prevent reductions in Medicare rates resulting from the Centers for Medicare and Medicaid Services (CMS) calculations of clinical services provided and their associated costs (Plested, 2006).

In 2015, MACRA (Medicare Access and CHIP Reauthorization Act) replaced the Medicare's SGR (Steinbrook, 2015). The intention of MACRA was clearly to manage health care resource utilization and control costs. However, this strategy has failed on both accords. The momentum for payment reform and the specific payment mechanisms notwithstanding, physicians are likely to advocate for Medicare payment updates that at least keep up with inflation and the cost of living. At some point, the cumulative effect of the new payment updates will not keep up with physician costs, unless the volume and cost of services substantially decrease, which is the same underlying issue as with the old payment updates under the SGR process. Each physician continues to act like the herdsman on the commons trying to maximize his own personal happiness (Cassel and Brennan 2007).

Market dynamics create a comparable "moral-agency" dilemma between physicians and the commercial health insurance industry. The managed care experiment in the U.S. (i.e., Health Maintenance Organizations-HMOs) from the 1980's led to physicians' mistrust regarding where healthcare savings from cost-control measures are directed and the significance of profit margins within the industry (Cassel and Brennan 2007). The lack of transparency, accountability, and truthfulness on the part of U.S. health insurance companies created the perfect storm for a major paradigm shift towards value-based models of care (e.g., patient-centered medical homes and accountable care organizations) in the 1990's.

Renewing the medical commons globally can strengthen clinician accountability without undermining the quality of medical services at the point of care (Cassel and Brennan 2007). Today, consumer-directed medical care and efforts to increase transparency and accountability about quality, access, cost and appropriate use of health care services enables patients to make informed decisions about their medical care. This should result in the better use of today's limited health care resources without compromising the positive health outcomes related to the care provided. For the most part, clinicians today do not have a sense of responsibility for health care resource utilization, nor do they feel as if they have any influence over health care resource use within the global healthcare delivery system.

Are there examples that exist in the global healthcare delivery system today advanced as models of ethically and empirically-driven health care stewardship? Holistic, integrated and comprehensive healthcare delivery systems that provide some of the necessary organizational structures and processes that can bring transparency, accountability, and truthfulness back to all healthcare stakeholders can renew the medical commons (Nasiri et al, 2019). For example, a group practice accepts the notion of group responsibility. All the physicians within the group practice accept responsibility for the overall health, wellness, and well-being of every patient within the population the practice serves. This dual responsibility for a population and the individual patient compels the clinicians within the practice to be accountable for quality care, patient safety, trust and truthfulness, cost efficiency, and accessibility (Enthoven and Tollen 2004). Multispecialty group practice supported with proper incentives offers the coordination of chronic and complex disease management infrastructure to support the use of evidence-based medical care, and the ability to afford, invest in, and implement clinical information technology (Crosson 2004; Nasiri et al, 2019). Again, all is done in the name of transparency and accountability for quality care, accessibility, and cost efficiency within the context of using the medical commons responsibly.

6.4 The Patient-Centered Medical Home: Managing the U.S. Medical Commons.

One example of managing the medical commons with a major paradigm shift to a multi-level organizational structure and process change is the establishment of the primary medical home for individual patients and populations (AHRQ, 2019; NCQA, 2020). First access to care, coordination of care, longitudinal care, and comprehensive care are just a few of the attributes that the primary medical home will provide that allows for health care stewardship to flourish within today's US health care delivery system.

The Patient-Centered Medical Home (PCMH) (AHRQ, 2019; NCQA, 2020) is a model of care delivery designed to improve patient outcomes by providing comprehensive, patient-centered, and coordinated care. At its core, the PCMH emphasizes the use of primary medical care as the central hub for patient care management, ensuring that all aspects of a patient's health, wellness, and well-being are addressed cohesively and efficiently. This approach is structured around five foundational functions and attributes including comprehensive care, patient-centered care, coordinated care, accessible services, and quality/safety of care.

Comprehensive care within the PCMH model involves a team-based approach to healthcare delivery, where a primary care physician leads a group of healthcare professionals to meet the majority of a patient's physical and mental health care needs. This team may include physicians, nurse practitioners, nurses, pharmacists, nutritionists, social workers, and other professionals. By offering a broad spectrum of services, the PCMH aims to provide holistic care that addresses proactive care, acute illness and injury care, and chronic condition care.

Patient-centered care is a hallmark of the PCMH, emphasizing the importance of understanding and respecting each patient's unique culture, preferences, needs and values. This attribute encourages an active partnership between patients and their healthcare providers, fostering an agency where individuals and their families are empowered to take a participatory role in their own medical care. The over-arching goal is to create a healthcare experience that is personalized and responsive to individual patient preferences, needs, and values, thus enhancing patient satisfaction and engagement.

Coordinated care in the PCMH model ensures that medical care is organized across all elements of the broader healthcare delivery system, including specialty care, hospitals, home health care, and community services. This function is crucial for preventing fragmentation of care, which can lead to process errors, inefficiencies, and poorer health outcomes. By maintaining clear communication channels and using interoperable health information systems and technology, the PCMH facilitates seamless transitions of care with seamless continuity across different settings and providers.

Accessibility is another foundational aspect of the PCMH, which strives to make healthcare services more accessible by reducing obstacles and barriers to care. This may include offering extended office hours, providing easier scheduling options, and using telehealth services and broad-band technology to outreach to a subset of patients who may have difficulty accessing traditional in-person office visits. Enhancing access to care is fundamental to improving overall health outcomes and ensuring that patients receive timely and appropriate medical care.

Finally, the PCMH model places a strong emphasis on quality and safety, continually monitoring and improving healthcare practices to achieve the best possible patient outcomes. This involves the use of empirically-driven, evidence-based medicine, clinical decision-support tools, performance measurement, and quality improvement initiatives. By focusing on quality and safety, the PCMH aims to provide high-value care that is both effective and efficient, ultimately leading to better health outcomes and reduced healthcare costs (AHRQ, 2019; NCQA, 2020).

6.5 Macroeconomics Impact on Healthcare Stewardship

Macroeconomics significantly impacts healthcare stewardship by influencing the availability and allocation of resources within the healthcare system (Saltman et al., 2000). Economic growth, for instance, affects government revenues and, consequently, public spending on healthcare. In periods of economic prosperity, governments typically have more funds to allocate toward healthcare infrastructure, technology, and services, enhancing overall healthcare quality and access. Conversely, during economic downturns, budget cuts and resource constraints can lead to reduced funding for healthcare, negatively affecting service delivery, staffing, and the availability of medical supplies.

Inflation, a foundational macroeconomic variable, also plays a critical role in healthcare stewardship. Rising inflation increases the cost of medical supplies, pharmaceuticals, and healthcare services. This can strain healthcare budgets, especially in open, complex, and adaptive healthcare systems reliant on fixed budgets or insurance reimbursements. High inflation rates can force healthcare providers to make difficult choices, such as reducing services or delaying investments in new technology and refreshed facilities, ultimately impacting patient quality of medical care and positive health outcomes. Additionally, inflation can erode the purchasing power of individuals, making healthcare less affordable for many and potentially increasing the number of uninsured or underinsured individuals.

Unemployment rates, another macroeconomic factor, influence healthcare stewardship by affecting insurance coverage and health outcomes. High unemployment often leads to a loss of employer-sponsored health insurance in countries that have this model of payment for healthcare services, increasing the reliance on essential public health services or leading to gaps in medical coverage. This shift can strain local public health agencies and reduce access to essential healthcare services. Moreover, unemployment is associated with higher levels of stress overload and untreated diseases of despair (e.g., depression, anxiety, and suicidal ideation) increasing the demand for healthcare services at a time when resources may be more limited.

Financial exchange rates and international trade policies also impact healthcare stewardship by affecting the cost and availability of imported medical supplies, equipment, and pharmaceuticals. Fluctuations in international exchange rates can lead to increased costs for imported goods, affecting the financial sustainability of healthcare providers. Trade policies that impose tariffs or restrictions on medical goods can further complicate medical supply chains, leading to shortages and higher costs for essential healthcare-related items. These economic factors necessitate careful management and strategic planning by healthcare leaders to ensure that high-quality care remains accessible and affordable, despite the significant external economic pressures.

6.6 Microeconomics Impact on Healthcare Stewardship

Microeconomics profoundly impacts healthcare stewardship through the dynamics of supply and demand, pricing, and the behavior of individual healthcare providers and consumers (Costa-Font et al., 2020). The principles of supply and demand determine the availability and cost of healthcare services. For instance, in areas with a high concentration of healthcare providers, competition can drive down prices and improve service quality, benefiting patients. Conversely, in regions with limited healthcare resources, providers may have greater pricing power, leading to higher costs and potentially reduced access to care. Understanding these market dynamics is essential for healthcare stewards to ensure resources are allocated efficiently and equitably.

Pricing strategies in healthcare are another critical aspect of microeconomics that affects stewardship. Healthcare providers must balance the need to cover costs and generate sufficient revenue with the goal of making services affordable for patients. This involves making decisions about pricing for different services, negotiating reimbursement rates with insurers, and managing costs. Effective pricing strategies can ensure that healthcare organizations remain financially viable while providing high-quality care. However, pricing that is too high can limit access for patients, particularly those without sufficient insurance coverage or financial resources, highlighting the need for careful and ethical pricing decisions in healthcare stewardship.

The behavior of individual healthcare providers and patients also shapes the microeconomic landscape of healthcare. Providers make decisions about the allocation of their time and resources, which impacts the availability and quality of care delivered. For example, the decision to adopt new medical technologies or new clinical practice guidelines may improve patient outcomes but also involves significant upfront investment and potential downside-risk. On the patient side, decisions about when and where to seek care, adherence to treatment plans, and willingness to pay for services affect demand and utilization patterns. Healthcare stewards must understand these behaviors to design interventions and healthy public policies that align incentives with desired health outcomes, ensuring that resources are used effectively and efficiently.

Market structures and competition within the healthcare sector further influence microeconomic outcomes and stewardship. In monopolistic or oligopolistic markets, a few providers may dominate, potentially leading to higher prices and less innovation. Conversely, in more competitive markets, the pressure to attract and retain patients can significantly drive improvements in service quality and efficiency. Healthcare stewards must navigate these market structures, advocating for healthy public policies that encourage friendly competition and eliminate anti-competitive practices. By fostering a competitive yet collaborative environment, stewards can help ensure that healthcare systems deliver high-quality, accessible, and affordable care to all patients.

6.7 The Healthcare Marketplace

Since the time of Adam Smith, the majority of economists have advocated the use of free markets as a mechanism of promoting efficiency (Bassiry et al., 1993). Smith himself theorized that in a free market society, individuals of their own volition pursue their own self-interests. In doing so, these independent individuals will be led to produce an allocation of resources that maximizes society's utility. There are certain conditions that require free markets to exercise efficient resource allocation. Above all, the first condition is *perfect competition*. This means that a market exists in which no individual buyer or seller can affect the price by his or her own actions. Without this condition, market distortions occur resulting in inefficient allocations of resources.

The contemporary Adam Smith's free market for health care efficiency contains many distortions, including the asymmetry of information between clinicians and patients, the clinician's dual role as patient advocate and independent business owner, the effect of insurance in reducing the apparent cost of health care goods and services to patients (the full cost of health care is not charged to patients, aka "price-wedge" distortion), tax subsidies that have a similar effect on consumers' decisions to purchase insurance, and monopoly power bestowed on certain professions and in some countries, health insurance plans, thereby limiting competition.

According to Wells et al. (2007), there were three classic efforts to describe, quantify, and propose solutions to the free market distortions in health care delivery: the Medical Commons, the Rand Health Insurance Experiment, and the Consumer Choice Health Plan.

Discussed at length earlier in this chapter, Hiatt (1975) challenged the medical profession to establish who was responsible for protecting the *medical commons*. Hiatt referred to an article by Harden (1968) in which he used the analogy between the practice of medicine and the practice of a group of herdsmen whose cattle share a common pasture to describe two important health care market distortions: the *price wedge* and *asymmetry of information.* Keeping the number of animals small in relation to the capacity of the pasture allows each herdsman to increase his holdings without detriment to the general welfare. However, when the pasture reaches near capacity, each animal has the potential to do irrevocable harm to the long-term sustainability of the system through overconsumption. Hiatt argued that the medical commons was fast approaching depletion as a consequence of free access to health care goods and services resulting in overconsumption (i.e., moral hazard) and increasing costs exacerbated by rapidly expanding, expensive technology of unproven effectiveness.

Hiatt (1975) proposed three types of solutions to this lack of health care stewardship: clinical evaluation of current medical practices (i.e., evidence-based medicine), setting a national health agenda, and expanding health care to include the social determinants of health with a focus on proactive health prevention, promotion, and protection.

The second classic effort to describe, quantify, and propose solutions to the free market distortions in health care delivery was a study performed by the Rand Corporation entitled, the Rand Health Insurance Experiment (aka, HIE). This experiment was one of the largest randomized social trials of its kind and occurred during the 1970s. The study asked three basic questions: How do demand-side incentives (e.g., cost-sharing) affect total health care utilization and expenditures? What is the effect of cost-sharing on individual health status? What are the consequences of cost-sharing on appropriateness and quality of care? The first two questions address the effect of the *price-wedge* distortion on expenditures and health and the last question partially addresses the asymmetry of information (Wells et al. 2007).

The HIE enrolled 7700 adults less than sixty-five years of age in six sites (both urban and rural) across the United States. Study participants were randomly assigned to one of five groups:

1. Free medical care
2. 25 percent cost-sharing up to a maximum amount
3. 50 percent cost-sharing up to a maximum amount

4. 95 percent cost-sharing up to a maximum amount
5. Enrollment in a nonprofit health maintenance organization

All participants enrolled in the study were observed for a period of three to five years.

The effects of cost sharing on total expenditures were significant. Patients with a 25 percent coinsurance spent 20 percent less than those with free medical care, and those with 95 percent coinsurance spent approximately 30 percent less (Newhouse et al. 1981). Savings resulted from using fewer health care services rather than from finding health care providers with lower prices as individual financial responsibility for health care increased by creating a self-imposed restriction on utilization of healthcare resources.

There was no significant difference in overall individual health status between study participants who enrolled in free medical care and cost-sharing insurance plans (Brook et al. 1983). Families that used medical resources more frequently because of less out-of-pocket costs gained no significant observed advantage in terms of their own individual health improvement. Smoking status, cholesterol level, and weight did not differ as a function of insurance plan. There was one significant change among the poorest and sickest 6 percent of the sample. Free medical care was associated with better health specifically related to diastolic blood pressure, functional far vision, and risk of dying. Increased physician visits by those families in the free medical care group resulted in the positive health care outcomes.

The effect of cost-sharing on appropriateness of care revealed that cost-sharing reduced overall medical care utilization. However, there was no significant difference in change of use between effective and ineffective care (Lohr et al. 1986). Restriction of health care as a result of facing a higher *true cost* did not result in patients increasing the proportion of their medical care considered appropriate by the Rand experts or conversely, decreasing the proportion considered inappropriate. Siu et al. (1986) reported no statistically significant differences in inappropriate hospital admissions between cost-sharing plans. According to Wells et al. (2007), these published findings from the Rand Corporation Health Insurance Experiment may have influenced the trend in health care delivery to focus on supply-sided instruments to improve on appropriateness and quality of care (i.e., evidence-based medicine, clinical practice guidelines).

Enthoven (1978) studied and reported the third and final classic effort to describe, quantify, and propose solutions to free market distortions in health care. Enthoven developed a proposal for a new national health insurance system known as the Consumer Choice Health Plan (CCHP). This proposal involved controlling costs through direct competition by health insurance plans for enrollees. The plans would be required to do the following: offer a clear, understandable, and detailed breakdown of enrollment costs and covered benefits to all individuals, permit individuals to compare plans allowing for informed purchasing decisions, and encourage competitively priced coverage.

A separate provision existed for eliminating the existing system of tax incentives for health insurance. Federal-state partnerships would offer subsidies so that individuals could buy health insurance coverage. In order to control cost escalation, provisions existed for tax credits to individuals who purchased efficient health insurance plans. These tax credits would allow individuals to keep the savings generated from choosing a less costly health insurance plan. To ensure equity between rich and poor, another provision existed for vouchers to help lower-income individuals and families participate. To prevent plans from limiting enrollment to individuals from whom they are more likely to profit, the CCHP included rules such as open enrollment (i.e., no age, gender, or preexisting health condition exclusions) and flat premiums based on market-area actuarial costs (i.e., community rating).

Unfortunately, Enthoven's CCHP proposal failed because it would have been difficult to reverse existing tax incentives in place at the time of the initial proposal. However, parts of this proposal became the catalyst for a managed competition movement found in the 1994 Clinton health plan proposal.

6.8 21st Century Healthcare Market Forces in the U.S. Healthcare Delivery System.

According to Wells et al. (2007), in the early 2000's, US healthcare policy initiatives to reduce health care market distortions focused on both the supply and demand side of the health care marketplace. Pay-for-performance (P4P) encouraged the symmetry of information between patients and clinicians. Health plans, both public and private, offered to clinicians and health care institutions additional reimbursement for the delivery of high quality and appropriate medical services. This process ensured that patients receive important medical care that was not sufficiently prioritized before this process was initiated (Rosenthal and Dudley 2007).

Another solution for free market distortion takes aim at *price-wedge* distortions. Consumer-directed health care (CDHC) plans are in reality a variation of the catastrophic health insurance plans of the twentieth century or the 95 percent cost-sharing intervention group within the Rand Corporation Health Insurance Experiment. The CDHC plans attack price-wedge distortions by offering less expensive premiums in combination with high deductibles. This ultimately places responsibility back with individual patients to decide which health care goods and services to purchase. However, CDHC plans did not meet equitable criteria because these plans tend to favor children and healthy individuals while women and sick individuals face escalating health care costs (Woolhandler and Himmelstein 2007).

In 2007, a report surfaced in the medical literature describing a value-based insurance design (Fendrick and Chernew 2007). This design was a variation of the CDHC plan differentiating between health care services that are effective and cost-efficient and those that are not. These plans would cover both effective and cost-efficient services with a lower deductible thus reserving a high deductible for the "less desirable" interventions. Value-based insurance design attempted to fix both the *price-wedge* and *asymmetry of information* distortions. It resembled a health care goods and services formulary that would likely require significant administrative overhead to manage. (Oberlander et al. 2001).

Few employer-sponsored health care plans in the United States used value-based purchasing strategies that include incentives and programs aimed at improving the quality and performance of medical care for enrollees (Rosenthal et al. 2007). According to Wells et al. (2007), it was time for all the stakeholders of the US health care delivery system to admit that it was impossible to develop healthy public policy initiatives that can correct market distortions and produce an efficient health care marketplace. A truly efficient health care marketplace is one in which value received per dollar spent is maximized and the distribution of health care services is both fair and equitable for the population as a whole (Porter, 2021). Unfortunately, for the US health care delivery system, both the political system and the social consensus was not clear about whether every person in society has a right to health care regardless of their ability to pay.

Fast forward to 2024. The healthcare marketplace is influenced by several key forces that significantly impact healthcare stewardship. One of the primary drivers is the shifting payer landscape. The profitability of government segments, particularly Medicare Advantage, is projected to outpace that of commercial segments. This shift is driven by increased Medicare Advantage penetration and a growing dual-eligible population, despite regulatory pressures and economic uncertainties affecting Medicaid enrollment (Patel and Singhal, 2024; Rao, 2024).

Economic conditions, such as inflation and recession risks, also play a critical role. While inflation is projected to decrease, ongoing financial pressures compel healthcare organizations to focus on cost containment and efficiency improvements. This includes increased outsourcing of administrative functions and leveraging technology for better operational management (Patel and Singhal, 2024; Corpart, 2024). Labor shortages and wage inflation continue to challenge healthcare systems, necessitating innovations in workforce management and the adoption of digital health solutions to maintain service levels and financial stability (Rao, 2024; Corpart, 2024) Pharmacy sector dynamics, including regulatory changes and the evolution of specialty pharmacies, are also significant. The specialty pharmacy segment is experiencing rapid growth due to high utilization and the introduction of advanced therapies. However, this growth is tempered by reimbursement pressures and the need for increased transparency in drug pricing under new legislative frameworks like the Inflation Reduction Act of 2022 (Rao, 2024).

Additionally, the healthcare market is seeing a rise in private equity involvement, with strategic acquisitions and public-to-private deals becoming more common. This trend is facilitated by lower valuations and the strategic need for healthcare organizations to diversify and enhance their capabilities amid financial constraints (Rao, 2024).

Overall, these marketplace forces require healthcare stewards to navigate a complex environment of economic pressures, regulatory changes, and technological advancements, all while striving to improve care quality and maintain financial health.

Professional responsibilities for the medical commons are inherent within the clinician-patient interaction. The clinical care for patients diminishes when there are insufficient health care resources for use by all patients. Rising health care costs threaten the care of many by increasing the number of uninsured patients, thus restricting access to the current formal U.S. health care delivery system. Inappropriate utilization and waste of the finite health care resources abounds. The ethical challenges of managing the medical commons require clinicians to link clinical care to cost of care.

Market-based and regulatory approaches place the welfare of patients in the hands of people other than their clinicians. The distribution of health care resources today occurs without regard for the caring, commitment, clinical acumen, and wisdom of experienced clinicians. Clinicians will be required to make an active decision about the approach they will take both individually and collectively regarding health care stewardship activities. It is apparent that clinician engagement in the medical commons, along with the engagement of all the stakeholders of the U.S. health care delivery system is the only approach that will ensure proper and appropriate allocation of health care resources leading to improved health care outcomes.

6.9 The Impact of Policy on Health Economics and Healthcare Stewardship.

Healthy public policies play a critical role in shaping the landscape of health economics, influencing how resources are distributed, healthcare services are delivered, and health outcomes are achieved. For example, policies that expand health insurance coverage, such as the Affordable Care Act (ACA) in the United States, aim to increase access to healthcare and reduce financial barriers for individuals (Maruthappu et al., 2013). These healthy public policies can lead to broader population health improvements, but they also have significant economic implications, including changes in healthcare spending, insurance premiums, and the financial performance of healthcare providers. Additionally, policies that promote cost-containment measures, such as value-based care and bundled payments, aim to improve the efficiency of healthcare delivery by incentivizing providers to focus on quality rather than quantity of care. The impact of these policies is assessed through economic evaluations, which help determine their effectiveness in improving health outcomes relative to their costs (Jones, 2018).

Healthcare stewardship is heavily influenced by public policies that govern the management and allocation of health resources (Turner et al., 2021). Policies that establish regulatory frameworks ensure that healthcare providers adhere to standards that promote patient safety, quality of care, and ethical practices. For instance, policies mandating transparency in healthcare pricing can empower consumers to make more informed decisions, potentially driving competition and reducing costs. Policies that support public health initiatives, such as vaccination programs and tobacco control measures, are essential components of stewardship, aiming to prevent disease and promote health at the population level. Furthermore, policies that invest in health information technology can enhance the efficiency and coordination of care by improving data sharing and communication among providers. Effective stewardship policies also focus on workforce development, ensuring that there are sufficient and appropriately trained healthcare professionals to meet the needs of the population. By guiding the strategic allocation of resources and setting priorities, policies help stewards create sustainable healthcare systems that are capable of adapting to emerging challenges and opportunities.

6.10 How do Healthy Public Policies affect Healthcare Stewardship?

Healthy public policies are essential for guiding the planning, development, and implementation of healthcare services and systems. Here are several examples of healthy public policies from different areas of healthcare.

1. *Public Health Policies.*
   a) *Vaccination Policies*: Mandating vaccinations for children and adults to prevent the spread of infectious diseases, such as measles, mumps, and influenza.
   b) *Smoking Bans*: Implementing bans on smoking in public places to reduce exposure to secondhand smoke and lower the incidence of smoking-related diseases.

2. *Health Coverage and Insurance.*
   a) *Universal Health Coverage*: Policies aimed at providing health insurance to all citizens, such as the National Health Service (NHS) in the UK or the Affordable Care Act (ACA) in the United States.

b) *Medicaid and Medicare*: Government-funded programs in the United States that provide health coverage to the aged, disabled, blind, and low-income individuals (Medicaid) and those aged 65 and older (Medicare).

3. *Health Equity Policies.*
   a) *Reducing Health Disparities*: Initiatives to address determinants of health, such as housing, food security, education, and income, to reduce health disparities among different populations.
   b) *Cultural Humility Training*: Requiring healthcare providers to undergo training in cultural humility to improve care for diverse populations considering their cultural preferences, needs, and values with respect and empathy.

4. *Environmental Health Policies.*
   a) *Clean Air Act*: Legislation aimed at reducing air pollution to improve public health by setting limits on emissions from industries and vehicles.
   b) *Water Quality Standards*: Regulations ensuring safe drinking water by setting limits on contaminants and requiring regular testing and reporting.

5. *Proactive Health Policies.*
   a) *Routine Screenings*: Guidelines recommending regular screenings for conditions like cancer, diabetes, and hypertension to detect and treat diseases early.
   b) *Nutritional Guidelines*: Policies promoting healthy eating, such as dietary guidelines and food labeling requirements, to combat obesity and related health issues.

6. *Healthcare Quality and Safety.*
    a) *Patient Safety Regulations*: Policies mandating practices to reduce medical errors, such as surgical checklists and infection control protocols.
    b) *Accreditation and Certification*: Requiring healthcare facilities to obtain accreditation from recognized bodies to ensure they meet certain quality and safety standards.

7. *Mental Health Policies.*
    a) *Mental Health Parity*: Laws requiring health insurance plans to provide equal coverage for mental health and substance use disorders as for physical health conditions.
    b) *Community Mental Health Services*: Policies promoting the development and funding of community-based mental health services to improve access and reduce stigma. E.g., Mobile Response and Stabilization Services, Emergency Departments exclusively for acute psychiatric conditions.

8. *Pharmaceutical and Drug Policies.*
    a) *Drug Approval Processes*: Regulations governing the approval of new drugs to ensure they are safe, efficacious and effective before reaching the market.
    b) *Prescription Drug Monitoring Programs (PDMPs):* most if not all states have their own systems to track the prescribing and dispensing of controlled substances to combat opioid misuse and abuse.
    c) *Price containment* for vulnerable and marginalized Americans so they don't have to choose between their life-saving medication and housing or food.

9. *Global Health Policies.*
    a) *International Health Regulations (IHR)*: A legally binding framework for countries to prevent and respond to public health risks that have the potential to cross borders.

   b) *Global Health Security Agenda (GHSA)*: An international effort to strengthen the world's ability to prevent, detect, and respond to infectious disease threats.

10. *Health Information Technology.*
    a) *Electronic Health Records (EHR) Incentives*: Policies providing financial incentives for healthcare providers to adopt and meaningfully use EHRs and data platforms (e.g., population health and SDOH platforms integrated with EHRs) to improve patient care.
    b) *Telehealth Regulations*: Guidelines and policies promoting the use of telehealth services to increase access to care, especially in rural and underserved areas.

These examples illustrate the broad scope of global healthy public policies and their critical role in shaping global healthcare delivery systems, improving public health outcomes, and exercising transparent and accountable healthcare stewardship.

6.11 References.

1. Agency for Healthcare Research and Quality. (2019). Patient-Centered Medical Home Resource Center.
2. Bassiry, G. R., Jones, M. (1993) Adam Smith and the ethics of contemporary capitalism. *Journal of Business Ethics* 12: 621-627.
3. Brook, R. H., Ware Jr., J.E., Rogers, W.H., et al. (1983) Does free care improve adults' health? Results from a randomized controlled trial. *N Engl J Med* 309 (23): 1426–1434.
4. Cassel, C. K., Brennan, T.E. (2007) Managing medical resources: Return to the commons? *JAMA* 297 (22): 2518–2521.
5. Corpart, G. (2024). The Face of Healthcare in 2024. Global Health Intelligence.

6. Costa-Font, J., Turati, G., Batinti, A. (2020) The political economy of health and healthcare: The rise of the patient citizen. Cambridge University Press.
7. Crosson, J. (2004) Panelist for: Delivery systems matter! Improving quality and efficiency in health care (transcript). Kaiser Permanente Institute for Health Policy and Health Affairs 1-Day conference, March 17, 2004. Washington DC.
8. Daniels, N. (1986) Why saying no to patients in the United States is so hard: Cost containment, justice, and provider autonomy. *N Engl J Med* 314: 1380–1383.
9. Enthoven, A. C. (1978) Consumer-choice health plan, II: A national-health-insurance proposal based on regulated competition in the private sector. *N Engl J Med* 298 (13): 709–720.
10. Enthoven, A. C., Tollen, L.A. (2004) *Toward a 21st century health system: The contributions and promise of a prepaid group practice.* San Francisco, California: Jossey-Bass.
11. Fendrick, A. M., Chernew, M.A. (2007) Value-based insurance design: A "clinically sensitive, fiscally responsible" approach to mitigate the adverse clinical effects of high-deductible consumer-directed health plans. *J Gen Intern Med* 22 (6): 890–891.
12. Hardin, G. (1968) The tragedy of the commons. *Science* 168; 162: 1243–1248.
13. Hiatt, H. H. (1975) Protecting the medical commons: Who is responsible? *N Engl J Med* 293: 235–241.
14. Jones, L. K. (2016) The roadmap to value-based care. *JAMA Neurology* 73(10): 1173-1174.
15. Lohr, K. N., R. H. Brook, Kamberg, C.J., et al. (1986) Effect of cost-sharing on use of medically effective and less effective care. *Med Care* 24 (9): S31–S38.
16. Maruthappu, M., Ologunde, R., Gunarajasingam, A. (2013) Is health care a right? Health reforms in the USA and their impact upon the concept of care. *Annals of Medicine and Surgery* 2(1): 15-17.

17. Nasiri, T., A. Takian, Yazdani. S. (2019) Stewardship in Health, Designing a Multi-Layer Meta Model: A Review Article. *Iranian Journal of Public Health* 48(4): 579–592.
18. National Committee for Quality Assurance. (2020). NCQA Patient-Centered Medical Home (PCMH). Retrieved from https://www.ncqa.org/programs/health-care-providers-practices/patient-centered-medical-home-pcmh/.
19. Newhouse, J. P. (1993) Insurance Experiment Group. *Free for all? Lessons from the RAND Health Experiment.* Cambridge, MA: Harvard University Press.
20. Newhouse, J. P., W. G. Manning, C. N. Morris, et al. (1981) Some interim results from a controlled trial of cost sharing in health insurance. *N Engl J Med* 305 (25): 1501–1507.
21. Oberlander, J., T. Marmor, Jacobs, L. (2001). Rationing medical care: Rhetoric and reality in the Oregon Health Plan. *CMAJ* 164 (11): 1583–1587.
22. OECD. (2024) OECD Health Statistics 2024.
23. Patel, N, Singhal S. (2024) What to Expect in US Healthcare in 2024 and beyond. McKinsey and Company.
24. Porter, M. E. (2010). What is value in health care? *New England Journal of Medicine* 363(26): 2477-2481.
25. Plested, W. G. (2006) Here we go again: Same SGR song, newest version. *AMNews.* October 16. http://www.ama-assn.org/amednews/2006/10/116/edca1016.
26. Rao, N. (2024) 2024 Healthcare Services Outlook: Challenges and Opportunities. McKinsey and Company.
27. Rosenthal, M. B., and R. A. Dudley. (2007) Pay-for-performance: Will the latest payment trend improve care? *JAMA.* 297 (7): 740–744.
28. Rosenthal, M. B., B. E. Landon, S. T. Normand, R. G. Frank, T. S. Ahmad, and A. M. Epstein. (2007) Employers' use of value-based purchasing strategies. *JAMA* 298 (19): 2281–2288.

29. Saltman, R. B., Ferroussier-Davis, O. (2000) The concept of stewardship in health policy. *Bulletin of the World Health Organization* 78(6): 732-739.
30. Siu, A.L., F. A. Sonnenberg, W. G. Manning, et al. (1986) Inappropriate use of hospitals in a randomized trial of health insurance plans. *N Engl J Med* 315 (20): 1259–1266.
31. Starr, P. (1982) *The social transformation of American medicine.* New York, New York: Basic Books, Inc.
32. Steinbrook, R. (2015) The Repeal of Medicare's Sustainable Growth Rate for Physician Payment. *JAMA* 313(20): 2025–2026.
33. Turner, H. C., Archer, R. A., Downey, et al. (2021) An introduction to the main types of economic evaluations used for informing priority setting and resource allocation in healthcare: key features, uses, and limitations. *Frontiers in Public Health 9*: 722927.
34. Wells, D. A., J. S. Ross, A. S. Detsky. (2007) What is different about the market for health care? *JAMA* 2007; 298 (23): 2785–2787.
35. Woolhandler, S., D. U. Himmelstein. (2007) Consumer-directed healthcare: Except for the healthy and wealthy it's unwise. *J Gen Intern Med* 22 (6): 879–881.

7.0

ESG Framework for a Sustainable Healthcare Marketplace.

In recent years, a transformational shift has emerged in the quest to redesign sustainable global healthcare systems—one rooted in the integration of environmental, social, and governance (ESG) principles. The ESG redesign framework serves as a roadmap for building a sustainable global healthcare marketplace that prioritizes resilience, equity, and ethical responsibility. By embedding ESG principles into healthcare system operations and finance, stakeholders are addressing systemic inefficiencies while fostering innovation and accountability.

The environmental dimension of the ESG framework calls for urgent action to reduce the healthcare sector's ecological footprint. Strategies presented include adopting renewable energy sources, implementing circular economy practices to minimize waste, and prioritizing sustainable procurement to ensure environmental responsibility. These approaches not only mitigate the sector's impact on the planet but also position healthcare delivery systems as leaders in the global sustainability movement.

Social equity and inclusivity serve as cornerstone elements of the ESG framework, promoting enhanced access to care for underserved populations. By fostering culturally competent practices and upholding ethical labor standards, we can shape a global healthcare delivery system that is both inclusive and equitable. Advancing health literacy and encouraging community engagement further strengthens this foundation. Addressing disparities in healthcare service delivery not only drives systemic improvement but also elevates societal well-being on a global scale.

Governance serves as the foundation of the ESG framework, ensuring ethical decision-making, transparency, and accountability throughout the global healthcare value chain. Prioritizing the alignment of organizational strategies with ESG principles is essential for fostering trust and strengthening resilience in healthcare delivery systems. Robust governance structures, effective stakeholder collaboration, and transparent reporting emerge as vital tools to achieve these goals.

Through the integration of ESG principles, the global healthcare marketplace can transform into a pioneering model of sustainability and innovation, equipped to tackle worldwide challenges with foresight and determination. This transformational approach stands as a vital global call to action, urging healthcare professionals, policymakers, and healthcare delivery systems to embrace the ESG framework as a cornerstone for reimagining healthcare services delivery around-the-globe. By doing so, it paves the way for a brighter, more equitable future for humanity and the planet alike.

7.1 Business Development and the Global Medical-Industrial Complex (GMIC).

The global medical-industrial complex (GMIC) is undergoing significant transformations driven by technological advancements, evolving patient expectations, and shifting regulatory landscapes.

One of the most prominent developments is the integration and interoperability of digital health technologies (Awad, 2021).

Telemedicine, wearable devices, and mobile health applications are revolutionizing patient healthcare services by providing remote monitoring, real-time data sharing, and personalized treatment plans. These innovations enhance patient engagement, reduce healthcare costs, and improve access to healthcare services, particularly in underserved areas.

Another critical development is the rise of value-based alternative models of care (e.g., accountable care organizations, accountable care communities, patient-centered medical homes), which shift the focus from volume to value. Instead of being compensated based on the number of services provided, healthcare providers around-the globe are rewarded for improving patient outcomes through high-quality care and reducing costs (Lewis, 2022). This approach encourages proactive care, chronic disease management, and coordinated care, leading to better health outcomes and more efficient use of limited healthcare resources. It also necessitates closer collaboration and cooperation between healthcare providers, payers, and patients, fostering a more integrated and comprehensive healthcare delivery system.

Artificial Intelligence (AI), machine learning (ML), and data analytics are also playing transformational roles in global healthcare services delivery (Scenario et al., 2021; Kahn et al., 2020). AI-powered tools assist in diagnostics, treatment recommendations, and predictive analytics, enabling more accurate and timely interventions. For instance, machine learning (ML) algorithms can analyze medical images with high precision, aiding in early detection of diseases such as cancer. Big data analytics helps identify trends and patterns in patient populations, informing public health strategies and personalized medicine initiatives. These technologies not only enhance clinical decision-making but also streamline administrative processes, reducing the burden on healthcare professionals to focus on the patient experience.

Personalized medicine is another burgeoning area, driven by advances in systems biology (e.g., genomics and other bio-molecular OMICs), computational science and mathematics, and big data analytics. (Hassan et al., 2022). By tailoring treatments to individual genetic profiles along with individual's epigenetic expression, healthcare providers can offer more effective and targeted therapies, minimizing adverse effects and improving patient outcomes. This approach is particularly impactful in the treatment of complex diseases such as cancer, where personalized therapies can significantly increase survival rates. The growth of biopharmaceuticals and precision medicine is also fostering partnerships between healthcare providers, academic and research institutions, and biotechnology companies, accelerating the development of personalized and innovative diagnostics and treatments.

Lastly, the focus on people-centered care is transforming the global healthcare landscape (Huffstetler et al., 2023). Today's patients demand more than conventional medical treatments; they seek primary care providers who embrace holistic, comprehensive, and integrated systems of health that address their physical, emotional, and social needs. In response, healthcare providers are enhancing patient communication, refining medical care environments, and offering supportive services such as mental health counseling and optimized scientific wellness programs. This shift toward person-centered care not only elevates patient experiences, satisfaction, and health outcomes but also fortifies the global healthcare delivery system by addressing the broader determinants of health, including social, economic, environmental, and commercial factors.

These contemporary advancements are collectively transforming global healthcare services delivery into a more efficient, effective, and people-centered system. By harnessing the power of digital health technologies, value-based care models, artificial intelligence, data analytics, personalized medicine, and holistic, integrated approaches to patient care, the global medical-industrial complex is redefining its capacity to address the evolving needs of all individuals and communities.

Digital health technologies, such as telemedicine, wearable devices, and remote monitoring tools, enhance accessibility and empower patients to actively engage in their health management. Value-based care models focus on delivering high-quality care by aligning provider incentives with patient outcomes, ensuring that limited healthcare resources are used efficiently and equitably. The integration of AI and data analytics enables precise diagnostics, predictive modeling, and tailored treatment plans, allowing for earlier interventions and improved health outcomes.

Personalized medicine marks a major paradigm shift by tailoring medical treatments to the unique genetic, environmental, and lifestyle factors of each individual. This approach not only optimizes therapeutic efficacy but also reduces unnecessary treatments, improving both the patient experience and cost efficiency. Additionally, holistic, integrated systems of health consider the physical, emotional, social, and environmental dimensions of health, emphasizing the importance of mental health support, community engagement, and environmental determinants of well-being.

Together, these innovations position the global healthcare delivery system for sustainable growth and continuous innovation, paving the way for a successful transformation towards Universal Health Coverage (UHC), Sustainable Development Goal (SDG) # 3, and the American healthcare delivery system focus on achieving the Quintuple Aim (Nundy et al., 2022). This framework aspires to achieve a positive experience for both patients and providers, superior health outcomes at both individual and population levels, and the elimination of health disparities, thereby advancing health equity. By addressing these critical dimensions, the GMIC not only enhances its operational resilience but also contributes to a more inclusive and equitable global health landscape.

7.2 Environmental Footprint of the Global Medical-Industrial Complex (GMIC).

The global medical-industrial complex (GMIC), while committed to improving health outcomes, has a substantial environmental footprint primarily resulting from its operational activities (Chung and Meltzer, 2009; Lenzen et al., 2020). Hospitals, healthcare facilities, and the entire medical supply chain are significant consumers of energy, water, and materials, and they generate a considerable amount of waste. The energy consumption in healthcare facilities is particularly high due to the need for continuous operation of medical equipment, HVAC systems, lighting, and other essential services. Hospitals, which operate 24/7, often require intensive energy use to maintain critical environments such as operating rooms and intensive care units. This high-energy demand contributes significantly to high volumes of greenhouse gas emissions, with healthcare facilities ranking among the highest energy consumers in the commercial sector.

A critical aspect of the GMIC's environmental footprint is its water usage. Healthcare facilities use vast quantities of water for various purposes, including sanitation, sterilization, and patient care. The reliance on water-intensive processes, such as laundering hospital linens, and operating cooling systems, adds to a hospital's overall environmental impact. Managing water usage efficiently and implementing water-saving technologies are essential steps to mitigate this impact (Peters et al., 2018). However, the challenge lies in balancing the need for rigorous hygiene and sanitation standards with the imperative to conserve water resources.

Waste generation in the GMIC is substantial and complex, comprising various types of waste, including general waste, hazardous waste, and medical waste. Medical waste, which includes items like used needles, contaminated dressings, and pharmaceutical products, poses significant disposal challenges due to its potential to spread infections and harmful substances. The proper handling and disposal of medical waste are critical to minimize environmental contamination and public health risks. Additionally, the widespread use of single-use medical supplies and the medical supply chain packaging contributes to the high volume of waste generated, further exacerbating the environmental burden (Windfield et al., 2015).

The GMIC also has a significant impact through its medical supply chain operations (Jacobs, 2018). The production, transportation, and disposal of medical supplies and pharmaceuticals involve considerable resource use and produces large amounts of GHG emissions. The procurement of medical equipment, pharmaceuticals, and other supplies significantly contributes to the production of GHG emissions and environmental degradation. By adopting sustainable procurement practices, such as sourcing from environmentally responsible suppliers and reducing the reliance on single-use items, the GMIC can mitigate the environmental impact of their supply chains.

Efforts to reduce the environmental footprint within the GMIC are gaining momentum. Many new healthcare facilities are adopting green building standards, such as LEED certification, to improve energy and water efficiency, reduce waste, and create healthier indoor environments (Sentman, 2009). Implementing energy-efficient technologies, such as LED lighting, energy management systems, and renewable energy sources, can significantly lower energy consumption and emissions. Waste reduction initiatives, such as recycling programs, reprocessing of medical devices, and the use of biodegradable materials, are also being explored to address the waste challenge. Water conservation measures, including the use of low-flow fixtures and water-efficient landscaping, help reduce water usage without compromising hygiene standards.

The environmental footprint of the GMIC operations is substantial, driven by high energy and water consumption, significant waste generation, and a complex medical supply chain. Addressing these environmental impacts requires a comprehensive approach that integrates sustainable practices into all aspects of GMIC operations (Lenzen et al., 2020). By adopting energy-efficient technologies, improving water management, reducing waste, and promoting sustainable procurement, global healthcare delivery systems can significantly reduce their environmental footprint while continuing to provide high-quality care at the lowest cost.

For example, Eckelman et al. (2020) summarizes the importance of holding US healthcare entities accountable for human disease as a public health imperative:

"... US health care activities contribute to substantial quantities of environmental emissions and disease burden, contrary to the mission to "First, do no harm." Health care organizations should take concrete steps to measure and reduce their carbon pollution. Mandated emissions reporting would inform science-based interventions and facilitate rapid adoption of sustainable health care practices that could dramatically reduce health care pollution and improve public health."

7.3 ESG Measures in the Global Medical-Industrial Complex (GMIC).

Environmental, Social, and Governance (ESG) measures play a pivotal role in ensuring that the GMIC, the medical supply chain, and the global healthcare delivery system function efficiently, sustainably, equitably, and ethically. By addressing the wide-ranging impacts of healthcare systems on the environment, society, and governance structures, these measures promote a holistic approach to service delivery. This comprehensive framework benefits all stakeholders, fostering resilience, accountability, and long-term environmental sustainability in global healthcare systems. (Lueng et al., 2023; Chime, 2024; Filipiak et al., 2021).

7.3.1 Environmental Measures

Environmental measures in the GMIC, the medical supply chain, and the global healthcare delivery system focus on reducing GHG emissions, encouraging the use of renewable energy, waste management, sustainable procurement, and water conservation. Key initiatives include:

1. *Energy Efficiency and Renewable Energy*: The GMIC can invest in energy-efficient technologies and renewable energy sources to reduce their GHG emissions. This includes upgrading to LED

lighting, using energy-efficient HVAC systems, and installing solar panels or wind turbines.

2. *Waste Management*: Effective waste management practices are essential in the GMIC to minimize environmental impact. This involves reducing, reusing, and recycling materials where possible, as well as proper disposal of hazardous medical waste. Implementing single-use plastics reduction programs and promoting the use of sustainable manufacturing materials also contribute to waste management efforts.
3. *Sustainable Procurement*: Adopting sustainable procurement practices throughout the medical supply chain ensures that the products and services used downstream by hospitals and healthcare facilities are environmentally friendly. This includes purchasing from the GMIC's medical supply chain who prioritize sustainability, opting for products with minimal packaging, and choosing reusable or recyclable medical supplies.
4. *Water Conservation*. The GMIC and medical supply chain can implement water-saving technologies and practices to reduce water consumption. This can involve installing low-flow fixtures, using water-efficient landscaping, and recycling water where appropriate.

7.3.2 Social Measures.

Social measures in the GMIC, the medical supply chain, and the global healthcare delivery system focus on promoting health equity, community engagement, employee wellness and well-being, and people-centered care. Key initiatives include:

1. *Health Equity*: Ensuring equitable access to healthcare services delivery for all populations, particularly marginalized and underserved communities. This involves addressing the determinants of health, such as safe housing, education, and income, and developing targeted programs to reduce health disparities.

2. *Community Engagement*: Building strong relationships with the community through outreach programs, partnerships with local organizations, and active involvement in community health initiatives. This engagement enables healthcare providers to cultivate cultural competence, gain a deeper understanding of community dynamics, and effectively address the unique health needs of the populations they serve.
3. *Employee Wellness and Well-being*. Promoting the wellness and well-being of healthcare workers by providing a safe and supportive work environment, offering professional development opportunities, and ensuring fair labor practices. Supporting mental health and work-life balance for employees is also foundational to recruitment and retention of a strong and professional healthcare workforce.
4. *People-Centered Care*: Ensuring that healthcare services delivery is people-centered, with a focus on improving the people experience and health outcomes. This involves providing culturally competent care, involving patients in shared decision-making, and offering education and support for self-management of health conditions.

7.3.3 Governance Measures

Governance measures in the GMIC, the medical supply chain, and the global healthcare delivery system focus on ensuring ethical leadership, transparency and accountability, compliance and risk management, and representation and belonging. Key initiatives include:

1. *Ethical Leadership*. Establishing strong leadership that prioritizes ethical decision-making and truthfulness. This involves setting clear ethical standards and using empirically-derived evidence-based health management ensuring that healthcare leaders at all levels adhere to these principles and practices.

2. *Transparency and Accountability*: Implementing transparent reporting practices and holding all stakeholders accountable for their actions. This includes regular disclosure of financial performance, environmental impact, and social initiatives, as well as engaging with stakeholders to gather voice of the customer and address concerns.
3. *Compliance and Risk Management*: Ensuring compliance with all relevant laws and regulations, and implementing robust risk management practices to identify, assess, and mitigate potential risks. This includes maintaining patient privacy and data security, adhering to clinical and safety standards, and addressing any unethical behavior promptly.
4. *Representation and Belonging*. Promoting representation and belonging within the healthcare workforce and leadership. This involves implementing policies and practices that foster an inclusive environment, supporting the recruitment and retention of diverse talent, and addressing any barriers to equal opportunity.

The integration of ESG measures into healthcare services delivery ensures that the GMIC, the medical supply chain, and the global healthcare delivery system function in a collaborative, responsible, and sustainable manner. This approach fosters improved health outcomes, empowers stronger communities, and supports a healthier planet. By prioritizing environmental, social, and governance considerations, these entities can establish a more resilient and equitable global healthcare delivery system designed to meet the needs of both present and future generations.

7.4 Basics of ESG Reporting in the Global Medical-industrial Complex (GMIC).

ESG reporting is a critical tool for the GMIC, medical supply chain, and the global healthcare delivery system, all of which are committed to environmental sustainability, transparency, and accountability. At its core, ESG reporting involves the systematic disclosure of information related to any healthcare entity's (e.g., hospital, health system, medical supply chain, GMIC) environmental, social, and governance practices. This process provides stakeholders—including investors, customers, employees, and regulators—with a clear understanding of a healthcare entity's efforts to manage ESG-related risks and opportunities. Effective ESG reporting starts with identifying the key issues that are most relevant to business operations and stakeholders. Conducting a materiality assessment allows healthcare entities to prioritize the most significant impacts and opportunities, ensuring that reports highlight the areas of greatest importance to their audience (Gabius, 2023; Arvidsson et al., 2022; Sellhorn et al., 2024).

Environmental reporting involves the disclosure of data related to a healthcare entity's environmental impact, including GHG emissions, energy consumption, water usage, waste management, and efforts to mitigate climate change. The environmental report should focus on the entity's strategies to reduce GHG emissions, enhance energy efficiency, and promote sustainable resource use. Key performance indicators might include total GHG emissions, percentage of energy from renewable sources, water consumption per unit of production, and waste recycling rates. These indicators provide a quantifiable measure of environmental performance and progress over time, demonstrating the healthcare entity's commitment to environmental sustainability.

Social reporting covers a broad range of topics related to a healthcare entity's impact on people and society. This includes labor practices, employee wellness and well-being, representation and belonging, community engagement, collective impact, social justice, and human rights. Information provided to stakeholders include workforce demographics, employee training and professional development programs, health and safety initiatives, and efforts to promote a diverse and inclusive workplace. Additionally, reporting on community involvement and philanthropic activities highlights the entity's contribution to social well-being. Key performance indicators such as employee retention rates, OSHA-recordable workplace injury rates, diversity ratios, and community investment levels offer insight into the company's social performance and its commitment to positive societal impact.

Governance reporting focuses on the systems and processes that ensures the healthcare entity is managed ethically and transparently. This includes information on board composition and diversity, executive compensation, shareholder rights and responsibilities, and risk management practices. Companies should disclose their governance structures, including the roles and responsibilities of the board of directors and the executive management team, and any corporate policies related to ethical conduct, anti-corruption, and compliance. Key performance indicators in governance reporting might include board diversity statistics, executive pay ratios, frequency of board meetings, ESG sub-committee reporting, and outcomes of shareholder votes. Robust governance reporting demonstrates a company's commitment to accountability and ethical management.

The process of ESG reporting for the GMIC, medical supply chain, and the global healthcare delivery system, is no different than any other industry and involves not only the collection and analysis of relevant data but also the communication of this information in a clear, accessible, and transparent manner. The GMIC, medical supply chain, and the global healthcare delivery system entities typically publish their own ESG reports annually, often as part of their broader sustainability or corporate responsibility reports. These reports should be guided by recognized reporting frameworks and standards, such as the Global Reporting Initiative (GRI) (Rimmel, 2020), the Sustainability Accounting Standards Board (SASB) (Schooley et al, 2015), and the Task Force on Climate-related Financial Disclosures (TCFD) (Eccles et al., 2019). Adhering to these frameworks ensures that the reports are comprehensive, comparable, and credible, enhancing trust and confidence among stakeholders.

Finally, ESG reporting is not just about compliance but about continuous improvement and stakeholder engagement (Sampson,

2023). All healthcare entities should use the insights gained from ESG reporting to inform their sustainability strategies and drive performance improvements. Regularly updating stakeholders on progress, challenges, and future goals fosters transparency and accountability. Engaging with stakeholders through surveys, consultations, and feedback mechanisms can provide valuable insights into their expectations and concerns, helping to align the entity's ESG efforts with stakeholder priorities. By embedding ESG reporting into their business strategy and operational/financial plan, healthcare entities of the GMIC, medical supply chain, and the global healthcare delivery system can enhance their sustainability performance, build stronger relationships with stakeholders, and create long-term value.

7.5 Essential Components for Incorporating ESG Principles into Strategies for a Sustainable Global Medical-Industrial Complex (GMIC).

The essential components for incorporating ESG principles into strategies for an environmentally sustainable GMIC, medical supply chain, and global healthcare delivery system rely on a few foundational elements (Dathe et al., 2024; Leung et al., 2023).

One of the primary elements is a robust commitment to environmental sustainability. This involves adopting green principles and practices such as reducing energy consumption, managing waste effectively, and minimizing GHG emissions. Healthcare entities can implement energy-efficient technologies, invest in renewable energy sources, and adopt sustainable building practices for new facilities. Additionally, reducing the use of harmful chemicals and promoting the use of recyclable materials in medical supplies can significantly lower an entity's environmental footprint. These practices not only contribute to global sustainability efforts but also help in reducing operational costs and enhancing the healthcare entity's reputation among environmentally conscious stakeholders.

Another critical foundational element is fostering social responsibility. This includes addressing all determinants of health (i.e., social, economic, environmental, and commercial), such as economic stability, education, food instability, toxic exposures, and access to healthcare services delivery. Healthcare entities should aim to provide equitable care to all populations, especially underserved and marginalized communities, with cultural humility and empathy. Enhancing workforce representation and belonging is also essential, as a diverse workforce can better meet the needs of diverse patient populations. Additionally, prioritizing employee wellness and well-being and ensuring fair labor practices contribute to a positive organizational culture and improved people care. By addressing these social aspects, healthcare entities can build stronger, healthier communities and foster greater trust, transparency, accountability, and loyalty among patients and employees.

Strong governance structures are equally vital in embedding ESG principles into healthcare strategies. This involves establishing clear policies and procedures to ensure ethical conduct, compliance with regulations, and transparency and accountability in decision-making processes. Healthcare entities should develop comprehensive governance frameworks that include regular audits, risk-management assessments, and stakeholder engagement mechanisms. Engaging stakeholders such as patients, employees, investors, and community members in decision-making processes ensures that the entity's strategies align with the preferences, needs, values, and expectations of its stakeholders. Effective governance also involves continuous monitoring and reporting of ESG performance, allowing entities to measure progress, identify areas for improvement, and maintain accountability.

Integrating ESG principles into healthcare strategies also requires a commitment to continuous innovation and improvement. This means staying abreast of technological advancements and adopting new tools and practices that support environmental sustainability goals. For instance, leveraging digital health technologies can reduce the need for physical infrastructure and travel, thereby lowering GHG emissions. Additionally, investing in research and development of environmentally sustainable medical practices and products can drive innovation and set new standards within the GMIC and medical supply chain. Continuous education and training for healthcare professionals on ESG issues and environmentally sustainable principles and practices are also crucial, ensuring that the workforce is equipped to support and drive the specific entity's sustainability initiatives.

In summary, the foundational elements of applying ESG principles in sustainable healthcare strategies include a strong commitment to environmental sustainability, fostering social responsibility, establishing robust governance structures, and a dedication to continuous innovation and improvement. By focusing on these elements, healthcare entities can create value not only for themselves but also for their patients, employees, and the broader community. This holistic and integrative approach ensures that environmental sustainability is embedded in the healthcare entity's core operations, leading to long-term success and resilience.

7.6 Materializing ESG Sustainability in the Global Medical-industrial Complex (GMIC).

Materializing ESG (Environmental, Social, and Governance) principles in the healthcare sector (i.e., GMIC) requires a strategic integration of these principles into core business operations and long-term planning (Chui et al., 2023; Rane et al., 2024; Leung et al., 2023).

For environmental sustainability, healthcare entities can implement energy-efficient technologies, reduce medical waste, and minimize their environmental footprint. This can be achieved by adopting renewable energy sources, optimizing water usage, and implementing green building standards for new facilities (Sentman, 2009; Vierra, 2016). For example, hospitals can invest in solar panels, energy-efficient lighting, and HVAC systems to reduce energy consumption. Additionally, reducing single-use plastics and properly managing medical waste can significantly decrease environmental impact. These practices not only contribute to global environmental sustainability goals but also lead to cost savings and enhanced organizational reputation.

Social responsibility in healthcare involves addressing all the determinants of health, promoting equity, and improving community health outcomes. Healthcare entities can work towards ensuring that all patients, regardless of socioeconomic status, have access to high-quality care. This can be done by expanding telehealth services, creating outreach programs for underserved populations, and partnering with community organizations to address social needs such as housing, nutrition, and education. Furthermore, fostering a diverse and inclusive workforce is essential. Healthcare providers can implement policies and practices that promote representation and belonging in hiring, provide cultural competency training, and support the career development and advancement of underrepresented groups. These efforts not only improve patient care but also enhance employee recruitment and retention and organizational culture.

Strong governance practices are crucial for embedding ESG principles into the business model of healthcare entities. This includes establishing transparent decision-making processes, robust risk management frameworks, and ethical business practices. Effective governance requires regular audits, compliance with regulations, and transparent reporting on ESG performance. By creating a governance structure that includes ESG goals, healthcare entities can ensure accountability and align their operations and financial holdings with stakeholder expectations. Engaging stakeholders such as patients, employees, investors, and community members in governance processes ensures that diverse perspectives are considered and that the entity's strategies are responsive to the needs, preferences, and expectations of those they serve.

To effectively materialize ESG principles, healthcare entities need to integrate them into their strategic planning and operational workflows. This involves setting clear ESG objectives, aligning them with business goals, and developing actionable plans to achieve them. For instance, entities can set targets for reducing GHG emissions, improving patient satisfaction scores, and increasing workforce diversity. These targets should be accompanied by detailed action plans, timelines, and assigned responsibilities. Regular monitoring and reporting on performance and progress towards these targets are essential to ensure accountability and continuous improvement. By embedding ESG into strategic planning, healthcare entities can drive sustainable growth and resilience.

Innovation and technology play a significant role in materializing ESG in the business of healthcare. Leveraging digital health technologies, such as telemedicine, electronic health records (EHRs), and AI-powered diagnostics, can enhance efficiency, reduce environmental impact, and improve health outcomes. For example, telemedicine reduces the need for physical visits, lowering the environmental footprint associated with patient travel. AI and data analytics can optimize resource use, predict patient needs, and personalize medical care, leading to better health outcomes and more efficient use of resources. Investing in research and development of sustainable evidence-based clinical practices and durable medical equipment also supports the advancement of ESG goals. By embracing innovation, healthcare entities can achieve sustainable business growth while adhering to ESG principles.

In summary, materializing ESG in the sustainable business of healthcare involves integrating environmental sustainability, social responsibility, and strong governance into core operations and strategic planning. This requires adopting green building practices, addressing all determinants of health, promoting equality and equity , and establishing robust governance frameworks. By setting clear ESG objectives, leveraging innovation, and engaging stakeholders, healthcare entities can create sustainable value for patients, employees, communities, and investors, ensuring long-term success and resilience.

7.7 A Roadmap for Implementing ESG Practices in the Global Medical-Industrial Complex (GMIC).

Implementing ESG practices throughout the Global Medical-Industrial Complex (GMIC), Medical Supply Chain, and Global Healthcare Delivery System requires a comprehensive and structured roadmap that outlines clear steps and objectives (Haar, 2024). The first step in this roadmap is to conduct a thorough materiality assessment of the current state of ESG practices within a healthcare entity. This involves evaluating existing policies, practices, procedures, and performance in environmental sustainability, social responsibility, and governance. Gathering data on energy usage, waste management, workforce diversity, patient satisfaction, the patient and clinical staff experience, and compliance with regulatory standards provides a baseline from which to measure progress. Engaging stakeholders, including employees, patients, investors, and community members, in this materiality assessment process ensures a holistic, integrated, and comprehensive understanding of the healthcare entity's strengths, weaknesses, opportunities and threats (Filipiak et al., 2021)

The next step is to develop a strategic plan that aligns ESG objectives with the healthcare entity's overall mission, vision, and strategic goals (Sepetis et al., 2024). This plan should include specific, measurable, achievable, relevant, and time-bound (SMART) goals for each of the ESG dimensions. For environmental sustainability, goals might include reducing GHG emissions, increasing energy efficiency, and minimizing medical waste. Social responsibility goals could focus on improving access and navigation to clinical care, enhancing workforce representation and belonging, and supporting community-based organizations and their community health initiatives. Governance goals might include strengthening compliance frameworks, increasing transparency, and improving stakeholder engagement. The strategic plan should also outline the resources required, including financial investments, technology, and personnel, to achieve these goals.

Once the strategic plan is in place, the next phase involves implementing targeted initiatives to achieve the outlined goals. For environmental sustainability, this could involve investing in energy-efficient technologies, implementing medical waste reduction programs, and adopting green building practices. Social initiatives might include expanding telehealth services to improve access to care, creating programs to support workforce diversity and inclusion, and partnering with community organizations to address all determinants of health. Governance initiatives could focus on enhancing compliance training for staff, establishing robust risk management frameworks, and improving transparency through regular ESG reporting. Each initiative should have a clear implementation timeline, assigned responsibilities, and key performance indicators for tracking progress.

Monitoring and evaluation are critical components of the roadmap, ensuring that the healthcare entity stays on track to meet its ESG goals. Establishing a robust monitoring system involves regularly collecting and analyzing data on key performance indicators related to environmental, social, and governance policies and practices. This data should be used to generate regular actionable reports, real-time dashboards, and annual scorecards with benchmarks that provide insights into progress, identify areas for improvement, and inform decision-making. Additionally, engaging stakeholders in the evaluation process helps maintain accountability and ensures that the entity's ESG efforts align with their expectations, preferences, needs and values.

The final step in the roadmap is continuous quality improvement and adaptation. ESG practices and standards are continually evolving, (e.g., annual changes to the global reporting initiative), and all healthcare entities must remain agile to anticipate and respond to these new challenges and opportunities in real time. This involves regularly reviewing and updating the strategic plan, incorporating feedback from stakeholders, and staying informed about emerging trends and best practices in ESG. Continuous education and development for staff on ESG issues are also essential, ensuring that the workforce remains knowledgeable and committed to the healthcare entity's environmental sustainability goals. By cultivating a culture of continuous quality improvement, healthcare organizations can maintain the effectiveness, relevance, and impact of their ESG practices over the long term.

A structured roadmap for implementing ESG practices in healthcare entities involves assessing the current state, developing a strategic plan with SMART goals, implementing targeted initiatives, monitoring and evaluating progress, and continuously improving and adapting strategies. By following this roadmap, healthcare entities can effectively integrate ESG principles into their daily operations, creating sustainable value for their patients, employees, communities, and stakeholders.

7.8 Value Creation for ESG in the Global Medical-Industrial Complex (GMIC).

The importance of Environmental, Social, and Governance (ESG) principles is growing in the GMIC, medical supply chain, and the global healthcare delivery system. ESG initiatives can lead to significant value creation by promoting environmental sustainability, enhancing social responsibility, and ensuring strong governance (Sherman et al, 2020). These principles not only align with the ethical imperatives of all healthcare entities but also drive financial performance, operational efficiency, and a positive patient experience.

From an environmental perspective, healthcare entities that adopt sustainable practices can reduce their environmental footprint and minimize medical waste, contributing to global efforts to combat climate change. Implementing energy-efficient systems, reducing single-use plastics, and promoting telehealth can lower operational costs and improve resource allocation. Additionally, environmentally conscious policies and practices can enhance a healthcare entity's reputation, attracting patients and investors who prioritize sustainability.

Social factors are equally crucial in value creation. Healthcare entities committed to social responsibility can improve community health outcomes by addressing all determinants of health. Fostering a diverse and inclusive workforce enhances innovation and patient care, as a variety of perspectives and experiences lead to better hard and soft skill development, critical thinking, problem solving, and decision-making. Furthermore, ensuring patient safety, privacy, and satisfaction builds trust, truthfulness, and loyalty, which are essential for long-term success.

Governance plays a pivotal role in maintaining the integrity and accountability of healthcare entities. Strong governance frameworks ensure compliance with regulations, ethical conduct, and transparent decision-making. By prioritizing governance, healthcare entities can mitigate risks, prevent fraud, and attract investment. Effective governance also involves stakeholder engagement, which includes patients, employees, investors, and the community, ensuring that the entity's strategies align with the preferences, needs, and value expectations of its stakeholders.

The integration of ESG principles and practices within the GMIC, the medical supply chain, and the global healthcare delivery system not only addresses environmental and social challenges but also enhances organizational performance and resilience. By committing to environmentally sustainable policies and practices, social responsibility, and robust governance, healthcare entities can create substantial value, benefiting both society and their triple-bottom line reporting of ESG (Crace et al., 2023; Chui et al., 2023; Leung et al., 2023.).

7.9 References.

1. Arvidsson, S., Dumay, J. (2022). Corporate ESG reporting quantity, quality and performance: Where to now for environmental policy and practice? *Business Strategy and the Environment* 31(3): 1091-1110.

2. Awad, A., Trenfield, S. J., Pollard, T. D., et al. (2021) Connected healthcare: Improving patient care using digital health technologies. *Advanced Drug Delivery Reviews* 178: 113958.
3. Chime, C. (2024) ESG Risks and Integration by Healthcare Companies.
4. Chiu, W. K., Fong, B. Y. F. (2023) Environmental, social and governance and sustainable development in healthcare. *Recent Trends of Research and Education in ESG and Sustainability*, 99-112.
5. Crace, L., Gehman, J. (2023) What really explains ESG performance? Disentangling the asymmetrical drivers of the triple bottom line. *Organization & Environment* 36(1): 150-178.
6. Chung, JW, Meltzer, DO. (2009) Estimate of the Carbon Footprint of the US Health Care Sector. JAMA 302(18):1970–1972.
7. Dathe, T., Helmold, M., Dathe, R., et al. (2024) *Implementing Environmental, Social and Governance (ESG) Principles for Sustainable Businesses: A Practical Guide in Sustainability Management.* Springer Nature.
8. Eccles, R. G., Krzus, M. P. (2019) Implementing the task force on climate-related financial disclosures recommendations: An assessment of corporate readiness. *Schmalenbach Business Review* 71: 287-293.
9. Eckelman, M. J., Huang, K., Lagasse, R., et.al. (2020). Health Care Pollution and Public Health Damage in The United States: An Update. *Health affairs (Project Hope)*, *39*(12), 2071–2079.
10. Filipiak, B. Z., Kiestrzyn, M. (2021) Potential ESG Risks in Entities of the Healthcare System. *In* Adapting and Mitigating Environmental, Social, and Governance Risk in Business (pp. 74-102). IGI Global.
11. Gabius, K. (2023) 8.1 Basics of Sustainability. *Sustainable Business Management* 129.

12. Haar, G. (2024). Developing a Sustainability Roadmap. In: The Great Transition to a Green and Circular Economy. Springer, Cham.
13. Hassan, M., Awan, F. M., Naz, A., et al. (2022) Innovations in genomics and big data analytics for personalized medicine and health care: A review. *International Journal of Molecular Sciences* 23(9): 4645.
14. Huffstetler, A. N., Phillips Jr, R. L., Leyns, et al. (2023). People-centered health services. *In* Person Centered Medicine (pp. 135-150). Cham: Springer International Publishing.
15. Jacobs, F. R., Chase, R. B. (2018). *Operations and supply chain management.* McGraw-Hill.
16. Khan, Z. F., Alotaibi, S. R. (2020) Applications of artificial intelligence and big data analytics in m-health: A healthcare system perspective. *Journal of Healthcare Engineering* 1: 8894694.
17. Lenzen, M., Malik, A., Li, M., et al. (2020) The environmental footprint of health care: a global assessment. *The Lancet Planetary Health* 4(7): e271-e279.
18. Lewis, S. (2022) Value-based healthcare: is it the way forward? *Future Healthcare Journal* 9(3): 211-215.
19. Leung, T. C. H., You, C. S. X. (2023). ESG Application in Sustainable Development of the Healthcare Industry. *In* Environmental, Social and Governance and Sustainable Development in Healthcare (pp. 47-64). Singapore: Springer Nature Singapore.
20. Nundy, S., Cooper, L. A., Mate, K. S. (2022) The quintuple aim for health care improvement: a new imperative to advance health equity. *JAMA* 327(6): 521-522.
21. Peters, A., Otter, J., Moldovan, A., et al. (2018) Keeping hospitals clean and safe without breaking the bank; Summary of the Healthcare Cleaning Forum 2018.
22. Rane, N., Choudhary, S., Rane, J. (2024). Artificial intelligence driven approaches to strengthening Environmental, Social, and Governance (ESG) criteria in sustainable business practices:

a review. *Social, and Governance (ESG) Criteria in Sustainable Business Practices: A review.*

23. Rimmel, G. (2020) Global reporting initiative. *In* Accounting for Sustainability (pp. 111-125). Routledge.
24. Sampson, C. J. (2023) Unlocking ESG in Healthcare: Governance Holds the Key. *Frontiers of Health Services Management* 39(3): 1-4.
25. Schooley, D. K., English, D. M. (2015) SASB: A pathway to sustainability reporting in the United States. *The CPA Journal* 85(4): 22.
26. Secinaro, S., Calandra, D., Secinaro, A., et al. (2021) The role of artificial intelligence in healthcare: a structured literature review. *BMC Medical Informatics and Decision Making* 21: 1-23.
27. Sellhorn, T., Wagner, V. (2024) The forces that shape mandatory ESG reporting. *In* Research Handbook on Environmental, Social and Corporate Governance (pp. 269-292). Edward Elgar Publishing.
28. Sentman, S. D. (2009). Healthy buildings: green building standards, benefits, and incentives. *The Journal of Biolaw and Business* 12(1): 4.
29. Sepetis, A., Rizos, F., Pierrakos, et al. (2024) A sustainable model for healthcare systems: The innovative approach of ESG and digital transformation. *Healthcare* 12(2): 156.
30. Sherman JD, Thiel C, MacNeill A, et al. (2020) The green print: advancement of environmental sustainability in health care. Resour Conserv Recycling. 161:104882.
31. Windfield, E. S., Brooks, M. S. L. (2015) Medical waste management–A review. *Journal of Environmental Management* 163: 98-108.
32. Vierra, S. (2016) Green building standards and certification systems. *National Institute of Building Sciences,* Washington, DC.

8.0

The Anthropocene Climate Change and Healthcare Stewardship.

The Anthropocene era, defined by significant human impact on the Earth's ecosystems and climate, presents unprecedented challenges to global health systems. Climate change, as a defining crisis of this epoch, exacerbates existing health inequities, disrupts healthcare services delivery, and intensifies the burden of disease. In response, healthcare stewardship emerges as a critical framework for addressing these challenges—emphasizing environmental sustainability, resilience, and equity in global healthcare delivery systems. This section explores how healthcare stewardship principles can guide the transformation of global healthcare delivery systems to adapt to and mitigate the impacts of the Anthropocene climate crisis, fostering a future where human and planetary health thrive in harmony.

8.1 Anthropocene Climate Change and Healthcare Stewardship.

Anthropocene (i.e., human-induced) climate change significantly impacts healthcare stewardship by exacerbating the frequency and severity of health crises, thereby straining healthcare resources and infrastructure (Barna et al., 2020, Shaw et al., 2021; Roka, 2020).

As climate change leads to more extreme weather events such as heatwaves, floods, and hurricanes, the incidence of climate-related health issues rises, including heat-related illnesses, respiratory problems from increased air pollution, and injuries from natural disasters (Tong et al., 2022). Global healthcare delivery systems must allocate additional resources to address these emergencies, often at the expense of other areas of clinical care. This increased demand challenges healthcare stewardship efforts to efficiently manage resources while maintaining high-quality care and preparedness for ongoing and future climate-related events.

Additionally, climate change influences the spread of infectious diseases, complicating healthcare stewardship (Morand et al. 2020). Warmer temperatures and changing precipitation patterns can expand the habitats of disease vectors like mosquitoes and ticks, leading to the spread of diseases such as malaria, dengue fever, and Lyme disease into previously unaffected regions. Global healthcare delivery systems must adapt by investing in disease surveillance, prevention, and treatment programs for these emerging infectious disease threats. This necessitates careful planning and resource allocation to prevent and manage outbreaks, emphasizing the importance of proactive and adaptable healthcare stewardship in a changing climate.

Climate change also exacerbates existing health disparities, making healthcare stewardship even more challenging (Sasser et al., 2018). Vulnerable populations, including the elderly, low-income communities, and those with preexisting health conditions, are disproportionately affected by climate-related health issues. These groups often have less access to healthcare and fewer resources to cope with climate impacts. Healthcare stewardship must prioritize equity by ensuring that limited healthcare resources are distributed fairly and that all populations have access to necessary care and support. This requires targeted interventions and healthy public policies aimed at reducing health disparities and improving resilience among the most vulnerable.

The environmental impacts of climate change necessitate a shift towards sustainable healthcare practices as part of responsible stewardship. The GMIC, medical supply chain, and global healthcare delivery systems are significant contributors to GHG emissions and environmental pollution. To mitigate their environmental footprint, all healthcare entities must adopt sustainable practices such as reducing energy consumption, minimizing medical waste, and sourcing eco-friendly products. These efforts not only contribute to global climate mitigation goals but also align with the principles of healthcare stewardship by promoting long-term environmental sustainability and essential public health services integration (Buse et al., 2018). Investing in green infrastructure and practices can also result in cost savings, which can be redirected to patient care and other critical operating expenses.

Finally, climate change underscores the importance of integrating climate resilience into healthcare stewardship strategies (Folke et al., 2020; . This involves building a resilient GMIC, medical supply chain, and global healthcare services delivery infrastructure that can withstand extreme weather events, ensuring reliable supply chains for medical supplies, and developing emergency preparedness and response plans that address climate-related existential health threats. By incorporating climate resilience into operational and financial planning, healthcare entities can better anticipate and respond to the challenges posed by a changing climate. This proactive approach is essential for maintaining the integrity and effectiveness of healthcare stewardship in the Anthropocene era, ensuring that all healthcare entities remain robust and capable of delivering high-quality care amidst environmental uncertainties (Kareiva et al., 2016).

8.2 Greenhouse Gas Emissions from the Global Medical-Industrial Complex (GMIC).

Greenhouse gas (GHG) emissions from the GMIC, medical supply chain, and global healthcare delivery system can be categorized into three scopes, as defined by the Greenhouse Gas Protocol (Feist, 2018; Ghaemi et al., 2020).

Scope 1 emissions are direct emissions from sources owned or controlled by the healthcare entity. These include emissions from on-site combustion of fossil fuels in boilers, furnaces, and generators, as well as emissions from entity-owned vehicles. For example, many hospitals around the world operate on-site power plants or emergency backup generators that rely on diesel or natural gas, contributing significantly to Scope 1 emissions. Addressing these emissions requires transitioning to cleaner energy sources, enhancing energy efficiency, and adopting low-emission vehicle fleets.

Scope 2 emissions encompass indirect GHG emissions from the consumption of purchased electricity, steam, heating, and cooling. These emissions occur at the entities where the energy is generated but are attributed to the healthcare entity that consumes the energy. Hospitals and healthcare facilities are typically large energy consumers due to their 24/7 operations and the need for high-powered medical equipment, extensive HVAC systems, and comprehensive lighting. To reduce Scope 2 emissions, healthcare entities should invest in energy-efficient infrastructure, participate in demand response programs, and purchase renewable energy credits or directly source renewable energy through on-site installations like solar panels or off-site renewable energy agreements.

Scope 3 emissions are the most extensive and encompass all other indirect emissions that occur in the value chain of the reporting healthcare entity. This includes GHG emissions from activities such as the production and transportation of purchased goods and services, business travel, employee commuting, waste disposal, and the use of sold products. In the healthcare sector, Scope 3 emissions can be substantial due to the complex GMIC and the medical and non-medical supply chains involved in manufacturing medical equipment, pharmaceuticals, and other healthcare products. Furthermore, the logistics of delivering supplies, managing medical waste, and even the emissions associated with patient and employee travel contribute significantly to Scope 3 emissions. Reducing these emissions involves collaborating, cooperating, and coordinating with suppliers to promote sustainable practices, optimizing logistics to minimize transportation emissions, and encouraging alternative commuting options for employees, such as telecommuting or public transportation incentives.

Efforts to mitigate GHG emissions across all three scopes require a comprehensive and coordinated approach (Davis et al., 2000). For Scope 1 and 2 emissions, healthcare entities can conduct energy audits to identify opportunities for efficiency improvements and retrofitting facilities with energy-efficient technologies. Implementing rigorous maintenance programs for heating and cooling systems and transitioning to low-carbon energy sources are also essential strategies. For Scope 3 emissions, engaging with suppliers to adopt greener practices, reducing the environmental footprint of procurement processes, and fostering a culture of sustainability within the organization are critical steps.

In conclusion, GHG emissions from the GMIC, the medical supply chain, and the global healthcare delivery system are significant and span across direct, indirect, and value chain-related activities, categorized into Scope 1, 2, and 3 emissions. By addressing these emissions through targeted strategies such as transitioning to renewable energy, enhancing energy efficiency, and promoting sustainable supply chain practices, all healthcare entities can substantially reduce their environmental impact. Integrating these efforts into a broader sustainability strategy not only helps mitigate Anthropocene climate change but also supports the healthcare entity's mission to protect and improve the public's health.

8.3 The Importance of Biodiversity Conservation in Healthcare Stewardship.

Biodiversity conservation play a critical role in healthcare stewardship by maintaining the balance of ecosystems, which directly impacts human health (WHO, 2021; Sharma et al., 2024).

Biodiversity refers to the variety of life on Earth, encompassing different species, genetic variations, and ecosystems. This diversity is essential for ecosystem resilience and the provision of ecosystem services, including clean air and water, fertile soil, and the regulation of climate. These services are foundational for all living things, especially human health, as they contribute to the availability of medicinal resources, nutritional security, and the prevention of infectious diseases.

Medicinal resources derived from biodiversity are indispensable to modern healthcare (Khasim et al., 2020). Many pharmaceuticals are either directly extracted from plants, animals, and microorganisms or synthesized based on compounds found in nature. For instance, the rosy periwinkle has been instrumental in developing treatments for leukemia and Hodgkin's disease. Similarly, the diverse microbial life found in soil has given rise to numerous antibiotics. The loss of biodiversity threatens the discovery of new medicinal resources and the continued availability of existing ones, underscoring the need for conservation efforts to safeguard these biological treasures.

Biodiversity conservation is also crucial for controlling disease vectors and maintaining ecosystem balances that prevent outbreaks (Lafferty, 2009). Healthy ecosystems regulate populations of disease-carrying organisms, such as mosquitoes, ticks, and rodents, thus reducing the incidence of diseases like malaria, Lyme disease, and hantavirus. Habitat destruction and climate change can disrupt these natural controls, leading to increased exposure to pathogens. By protecting and restoring natural habitats, conservation efforts help maintain these regulatory functions, thereby protecting human populations from emerging infectious diseases.

Furthermore, biodiversity conservation contributes to nutritional security, which is a cornerstone behind the production of positive health and optimization of wellness. (Iyiola et al., 2022). Diverse diets rich in various plant and animal species provide essential nutrients that are vital for human growth, development, and immune function. The erosion of genetic diversity in crops and livestock reduces the resilience of food systems to pests, diseases, and climate change, potentially leading to malnutrition and food insecurity. Conservation practices that promote agricultural biodiversity ensure the stability and sustainability of food supplies, supporting overall health, wellness, well-being, and resilience.

The stewardship of biodiversity conservation is vital for global healthcare services delivery. It supports the discovery and development of medicines, controls disease vectors, maintains ecosystem balances, and ensures nutritional security. Protecting biodiversity via conservation efforts is not only an environmental imperative but also a fundamental component of essential public health strategy and function.

8.4 Global Healthcare Delivery System ESG Response to Anthropocene Climate Change.

Environmental, Social, and Governance (ESG) principles are gaining heightened significance in the era of Anthropocene climate change, particularly within the realm of global healthcare delivery systems. Integrating ESG principles into healthcare operations promotes environmental sustainability, fosters social responsibility, and ensures ethical governance—providing a holistic framework to tackle the complex challenges brought about by climate change. (Lueng et al, 2023; Nakielski, 2023).

From an environmental perspective, global healthcare delivery systems can significantly reduce their environmental footprint by adopting green building standards, improving energy efficiency, and investing in renewable energy sources (Mostepaniuk et al., 2023). By minimizing medical waste through robust recycling programs and sustainable procurement practices, hospitals and healthcare facilities can lessen their environmental impact, contributing to broader climate change mitigation efforts.

On the social front, ESG principles advocate for fairness and belonging within healthcare delivery systems, crucial for addressing the health disparities exacerbated by climate change (NASEM, 2019). Vulnerable populations, such as low-income communities, the elderly, and those with chronic health conditions, often bear the brunt of climate-related health impacts. Global healthcare delivery systems can incorporate ESG frameworks to ensure equitable access to care and resources, developing targeted programs that enhance the resilience of these at-risk groups. By fostering community engagement and partnerships, healthcare entities can better understand and address the unique needs of diverse populations, promoting social justice and health equity.

Governance is another critical component of ESG in response to climate change, emphasizing transparency, accountability, and ethical decision-making within global healthcare delivery systems (Churet et al., 2014). Strong governance structures can ensure that sustainability initiatives are effectively implemented and that environmental and social goals are met. This involves setting clear targets for reducing GHG emissions, monitoring progress, and publicly reporting on health outcomes. Additionally, healthcare organizations must prioritize ethical considerations in their operations, such as ensuring workforce development, fair labor practices, protecting patient privacy, and maintaining high standards of clinical care. By adhering to robust governance principles, healthcare delivery systems can build trust with stakeholders and enhance their overall resilience to climate-related challenges.

Furthermore, integrating ESG criteria into global healthcare delivery systems can drive innovation and long-term financial performance (Sepetis et al., 2024). Sustainable practices often lead to cost savings, such as reduced energy expenditures and waste disposal costs, which can be reinvested into patient care and facility improvements. ESG-focused healthcare entities may also attract investment from socially responsible investors who prioritize environmental sustainability and ethical governance. This financial support can fund further advancements in advanced medical technology, infrastructure, and services, ultimately enhancing the quality and accessibility of healthcare. By aligning financial goals with ESG principles, healthcare delivery systems can achieve a balance between profitability and social responsibility (Bhattacharya et al., 2023).

Incorporating ESG principles into global healthcare delivery systems is essential for adapting to and mitigating the impacts of Anthropocene climate change. It requires a holistic, integrated, and comprehensive approach that considers environmental sustainability, social responsibility and fairness, and strong leadership through governance. By embracing these principles, healthcare entities can not only improve their operational efficiency and financial performance but also contribute to the broader societal effort to combat climate change. This proactive stance ensures that healthcare delivery systems are not only resilient and adaptable but also leaders in the global movement toward an all-encompassing environmentally sustainable and equitable future.

8.5 References

1. Barna, S., Maric, F., Simons, J., et al. (2020) Education for the Anthropocene: Planetary health, sustainable health care, and the health workforce. *Medical Teacher* 42(10): 1091-1096.
2. Bhattacharya, A., Bhattacharya, S. (2023) Integrating ESG pillars for business model innovation in the biopharmaceutical industry. *Australasian Accounting, Business and Finance Journal* 17(1): 127-150.
3. Buse, C. G., Oestreicher, J. S., Ellis, N. R., et al. (2018) Public health guide to field developments linking ecosystems, environ-

ments and health in the Anthropocene. *J Epidemiol Community Health* 72(5): 420-425.

4. Churet, C., Eccles, R. G. (2014) Integrated reporting, quality of management, and financial performance. *Journal of Applied Corporate Finance* 26(1): 56-64.
5. Davis, D., Krupnick, A., Thurston, G. (2000) The ancillary health benefits and costs of GHG mitigation: scope, scale, and credibility. *In* Ancillary Benefits and Costs of Greenhouse Gas Mitigation: Proceedings of an IPCC Co-sponsored Workshop (pp. 135-190).
6. Feist, T. (2018) The Three Scopes of Greenhouse Gas Emissions. *Journal of Property Management* 83(3): 24-26.
7. Folke, C., Carpenter, S. R., Chapin III, F. S., et al. (2020) Our Future in the Anthropocene Biosphere: Global sustainability and resilient societies.
8. Ghaemi, Z., Smith, A. D. (2020) A review on the quantification of life cycle greenhouse gas emissions at urban scale. *Journal of Cleaner Production* 252: 119634.
9. Iyiola, A. O., Babafemi, O. P., Ogundahunsi, O. E., et al. (2022) Food security: a pathway towards improved nutrition and biodiversity conservation. *In* Biodiversity in Africa: Potentials, Threats and Conservation (pp. 79-107). Singapore: Springer Nature Singapore.
10. Kareiva, P., Fuller, E. (2016) Beyond resilience: how to better prepare for the profound disruption of the Anthropocene. *Global Policy* 7: 107-118.
11. Khasim, S. M., Long, C., Thammasiri, K., et al. (Eds.). (2020) *Medicinal plants: biodiversity, sustainable utilization and conservation*. Singapore: Springer.
12. Lafferty, K. D. (2009) The ecology of climate change and infectious diseases. *Ecology* 90(4): 888-900.
13. Leung, T. C. H., You, C. S. X. (2023) ESG Application in Sustainable Development of the Healthcare Industry. *In* Environmental, Social and Governance and Sustainable

Development in Healthcare (pp. 47-64). Singapore: Springer Nature Singapore.

14. McLean, M., Behrens, G., Chase, et al. (2022) The medical education planetary health journey: advancing the agenda in the health professions requires eco-ethical leadership and inclusive collaboration. *Challenges* 13(2): 62.
15. Morand, S., Walther, B. A. (2020) The accelerated infectious disease risk in the Anthropocene: more outbreaks and wider global spread. *BioR*.
16. Mostepaniuk, A., Akalin, T., Parish, M. R. (2023) Practices pursuing the sustainability of a healthcare organization: A systematic review. *Sustainability* 15(3): 2353.
17. National Academies of Sciences, Engineering, and Medicine Division, & Committee on Integrating Social Needs Care into the Delivery of Health Care to Improve the Nation's Health. (2019) Integrating social care into the delivery of health care: Moving upstream to improve the nation's health.
18. Nakielski, M. L. (2023) Moving Forward with ESG, Sustainability, and Corporate Responsibility. *Frontiers of Health Services Management* 40(1): 33-39.
19. Roka, K. (2020) Anthropocene and climate change. *Climate Action* 20-32.
20. Sasser, J. S. (2018) Public health in the anthropocene: Exploring population fears and climate threats. *In* Global Health and Security (pp. 166-178). Routledge.
21. Sepetis, A., Rizos, F., Pierrakos, G., et al. (2024) A sustainable model for healthcare systems: The innovative approach of ESG and digital transformation. *Healthcare* 12(2): 156.
22. Sharma, I., Birman, S. (2024) Biodiversity Loss, Ecosystem Services, and Their Role in Promoting Sustainable Health. *In* The Climate-Health-Sustainability Nexus: Understanding the Interconnected Impact on Populations and the Environment (pp. 163-188). Cham: Springer Nature Switzerland.

23. Shaw, E., Walpole, S., McLean, M., et al. (2021) AMEE Consensus Statement: Planetary health and education for sustainable healthcare. *Medical Teacher* 43(3): 272-286.
24. Tong, S., Bambrick, H., Beggs, P. J., et al. (2022) Current and future threats to human health in the Anthropocene. *Environment international* 158: 106892.
25. World Health Organization. (2021) Nature, biodiversity and health: an overview of interconnections.

9.0

Anti-Microbial Stewardship: A Multi-level Global *Burning Platform* in Healthcare Stewardship and Sustainability

9.1 The *Burning Platform* in the Context of Healthcare

The term *burning platform* is a metaphor used in business to describe a situation where someone must make a drastic change in their behavior to avoid dire consequences. It comes from the true story of Andy Mochan, a worker on the Piper Alpha oil rig in the North Sea when it exploded in July 1988 (Conner, 1992). The explosion caused a massive fire that killed 167 men, and Mochan was awakened and faced a choice to either stay on the platform and burn alive or jump into the freezing water. Mochan chose to jump and survived, and later said that the experience caused a radical change in his behavior. In business, the concept of a *burning platform* is used to help people see the consequences of not changing and motivate them to embrace change. However, some say that overuse of the term can dilute the alignment needed to succeed, and that people can become numb and dismissive if there is a constant series of *burning platforms*.

In the general context of healthcare management (Warner et al., 2015), the *burning platform* refers to the critical and pressing issues that necessitate immediate and decisive action to prevent catastrophic health outcomes.

One critical *burning platform* in global healthcare services delivery is the chronic shortage of experienced healthcare professionals, significantly exacerbated by the recent COVID-19 pandemic (Teixeria, 2020). This shortage during the pandemic created a cascade of problems, including increased workloads for existing staff, potential compromises in patient care quality and safety, and higher operational costs due to the reliance on temporary staffing solutions. The stress and burnout experienced by healthcare workers worldwide further amplified the crisis, leading to higher turnover rates and a continuous cycle of staffing challenges. Addressing this burning platform required a multilayered and interdisciplinary approach, including strategic workforce development, improved working conditions, investment in education and training programs, and policies that support the retention and well-being of healthcare professionals.

The financial sustainability of global healthcare delivery systems is also a burning platform that demands urgent attention (Patel et al., 2019). Rising healthcare costs, driven by factors such as an aging population, advancements in medical technology, and the increasing prevalence of chronic diseases, place immense pressure on healthcare budgets. This financial strain can lead to difficult choices about resource allocation and may compromise the ability to provide high-quality care to all patients. Implementing value-based models of care that focus on delivering high-quality outcomes at lower costs is essential to address this burning platform. These care models emphasize the integration of holistic and comprehensive evidence-based medical care, patient engagement, and the use of data to drive efficient and effective healthcare delivery to best clinical practices.

Healthcare disparities and inequities represent yet another burning platform that calls for immediate action (Dickman et al., 2017). Upstream socioeconomic factors, geographic location, race, and ethnicity significantly influence access and navigation to healthcare services. These disparities have been bluntly highlighted by the COVID-19 pandemic, which disproportionately affected marginalized and underserved communities. To address this burning platform, global healthcare delivery systems must adopt strategies that promote equity, such as expanding access to care where people live, play and work, implementing culturally competent clinicians and *bricks and mortar* sites of care, and addressing the upstream determinants of health. Ensuring that all individuals receive just and equitable treatment is not only a moral and ethical imperative but also essential for improving the public's health, wellness, well-being, and resilience.

The burning platforms in the context of healthcare delivery systems and healthcare services delivery represent imperative matters that require immediate and decisive action to prevent catastrophic outcomes. Whether it is the healthcare workforce crisis, financial sustainability, or healthcare disparities, each of these challenges demands a strategic and coordinated response. By recognizing and addressing these burning platforms, global healthcare delivery systems can safeguard the public's health, ensure the delivery of high-quality care, and build a more resilient and equitable healthcare infrastructure for the future.

9.2 Anti-Microbial Stewardship: The Next *Burning Platform*?

Antimicrobial stewardship is considered a burning platform due to the immediate and severe threat posed by antibiotic resistance, which endangers global public health and the efficacy of modern medicine (Birgand et al., 2023; Frieri et al., 2017). The rapid emergence and spread of multidrug-resistant organisms significantly limit treatment options for infections, making once easily treatable illnesses potentially lethal. The urgency to address this issue cannot be overstated, as the continued misuse and overuse of antibiotics accelerate the development of resistance, leading to a scenario where common infections could become untreatable, and routine surgeries and medical procedures carry a higher risk of complications and mortality.

The economic impact of antibiotic resistance further underscores its status as a burning platform (Ahmad et al., 2019). Treating infections caused by resistant organisms is substantially more expensive than treating those caused by susceptible strains, due to the need for more complex, prolonged, and costly treatments. This financial burden affects not only individual patients but also global healthcare delivery systems and economies worldwide. Increased healthcare costs, longer hospital stays, and the need for more intensive care contribute to a strain on limited healthcare resources, making it imperative to implement effective antimicrobial stewardship programs to curb the economic drain caused by antibiotic resistance.

Patient safety is also at the heart of why antimicrobial stewardship is a burning platform (Domer et al., 2021). Inappropriate use of antibiotics can lead to adverse drug reactions, including life-threatening conditions such as *Clostridium difficile* infections, which often result from disrupted gut microbiota. These infections are difficult to treat and can cause severe morbidity and mortality, particularly in vulnerable populations such as the elderly and immunocompromised patients. Ensuring the appropriate use of antibiotics through stewardship programs helps prevent these adverse outcomes, safeguarding patient health and improving overall clinical outcomes.

Moreover, the global nature of antibiotic resistance exacerbates its urgency (Aljeldah, 2022). Resistant pathogens do not respect borders, and the misuse of antibiotics in one part of the world can have far-reaching consequences, spreading resistance internationally. This interconnectedness necessitates a coordinated and comprehensive response, involving international collaboration and adherence to stewardship principles across all healthcare settings world-wide. The failure to address antibiotic resistance on a global scale could lead to a public health crisis of unprecedented proportions, similar to the SARS-CoV-2 viral pandemic (i.e., COVID-19) of 2020, making antimicrobial stewardship an immediate priority.

Antimicrobial stewardship is considered a burning platform due to the calamitous and immediate consequences of international, national, and local antibiotic resistance for public health, economic stability, patient safety, and healthcare delivery systems. The urgent need to preserve the effectiveness of existing antibiotics and develop sustainable practices for their use underscores the critical importance of robust stewardship programs. Addressing this burning platform through concerted efforts can help prevent a future where common infections become deadly and modern medical practices are severely compromised.

9.3 Foundational Considerations for Global Anti-Microbial Stewardship.

Antimicrobial stewardship serves as a cornerstone of global healthcare delivery systems for several critical reasons. It plays a vital role in combating antibiotic resistance, ensuring the provision of effective patient care, and safeguarding public health, wellness, and overall well-being (Logan et al., 2019; Hwang et al., 2021).

Antibiotic resistance is one of the most pressing public health threats globally (Ferri et al., 2017; Dhingra et al., 2020). Misuse and overuse of antibiotics accelerate the development of resistant bacteria, rendering standard treatments ineffective. This leads to infections that are harder to treat, increasing morbidity, mortality, and healthcare costs. Antimicrobial stewardship aims to optimize antibiotic use to slow the emergence of resistance, ensuring that these life-saving drugs remain effective for future generations.

Effective antimicrobial stewardship ensures that patients receive the right antibiotic, at the right dose, for the right duration (Bankar et al., 2022). This targeted approach reduces the likelihood of adverse drug reactions and secondary infections, such as *Clostridium difficile,* which can be caused by broad-spectrum antibiotic use. By promoting precise and appropriate antibiotic use, stewardship programs enhance patient safety and improve overall clinical outcomes.

Infections caused by resistant organisms are more challenging and expensive to treat (Laxminarayan et al., 2016; Sipahi, 2008). They often require longer hospital stays, more intensive care, and more expensive and toxic drugs. By preventing the development and spread of resistance, antimicrobial stewardship helps contain healthcare costs. Additionally, reducing unnecessary antibiotic use minimizes the risk of costly adverse drug events and secondary infections.

The development of new antibiotics is slow and expensive, and the pipeline for new drugs is limited (Ardal et al., 2020). Stewardship programs play a critical role in preserving the effectiveness of existing antibiotics by promoting their judicious use. This is essential for maintaining the ability to treat bacterial infections effectively and for ensuring that antibiotics remain a viable option in the future.

Antimicrobial stewardship works hand in hand with infection prevention and control measures (Giamarellou et al., 2023). By reducing unnecessary antibiotic use, stewardship programs help decrease the prevalence of resistant bacteria within healthcare settings. This, in turn, reduces the risk of hospital-acquired infections and the spread of resistant organisms to vulnerable patient populations.

Antimicrobial stewardship is a cornerstone of essential public health functions and services to combat antibiotic resistance (Hwang et al., 2021; Tamma et al., 2011; Owens, 2008). National and international health organizations, such as the Centers for Disease Control and Prevention (CDC) and the World Health Organization (WHO), emphasize the importance of stewardship programs in their essential public health strategies, services, and functions to address resistance. These programs align with broader public health goals to improve healthcare quality and safety, protect vulnerable populations, and promote sustainable antibiotic use practices.

Stewardship programs play a vital role in educating healthcare providers about appropriate antibiotic prescribing practices and the importance of resistance. They also help inform patients about the proper use of antibiotics and the risks associated with misuse. This education fosters a culture of responsible antibiotic use and enhances the overall effectiveness of antimicrobial stewardship efforts (Lim et al., 2020; Ancillotti et al., 2018).

Antimicrobial stewardship is essential for addressing the critical threat of antibiotic resistance, improving patient outcomes, reducing healthcare costs, preserving the efficacy of current antibiotics, enhancing infection control, supporting public health initiatives, and educating both healthcare providers and patients. Through multi-layered and interdisciplinary efforts, stewardship programs play a pivotal role in safeguarding the public's health, wellness, and well-being and by actively pursuing the sustainable use of antibiotics.

9.4 Challenges to Implementing Anti-Microbial Stewardship into Global Healthcare Delivery Systems.

Implementing effective antimicrobial stewardship programs in global healthcare delivery systems presents several significant challenges. One of the primary obstacles is the variability in resources and infrastructure across different healthcare settings (Vaughn et al., 2024). Smaller hospitals and rural clinics often lack the necessary personnel, technology, and funding to establish and maintain robust stewardship programs. This disparity creates an uneven landscape where the quality of antibiotic management can vary widely, potentially undermining broader efforts to combat antibiotic resistance.

Another challenge lies in the complexity of changing established prescribing behaviors among healthcare providers (Powers, 2024). Many physicians and clinicians have long-standing habits and practices that may not align with current stewardship guidelines. Overcoming these ingrained behaviors requires continuous education, training, and the implementation of effective clinical decision support tools. However, ensuring consistent adherence to these guidelines across diverse and busy healthcare environments is difficult, especially when immediate clinical pressures and patient demands can influence decision-making.

The need for accurate and timely data is also a significant hurdle (Howard et al., 2024). Effective antimicrobial stewardship relies heavily on a learning health system framework for real-time data regarding antibiotic prescribing patterns, resistance trends, and patient outcomes. Many healthcare facilities struggle with integrating comprehensive data collection and analytic systems into their workflows. In the absence of valid and reliable data, it becomes challenging to monitor and assess the effectiveness of stewardship interventions, hindering the ability to pinpoint areas for improvement and measure their success.

Alignment between antimicrobial stewardship goals and broader healthcare delivery system priorities is often lacking (Ohlsen et al., 2024). Healthcare delivery systems are frequently under pressure to achieve financial targets, meet regulatory requirements, and manage patient throughput. In such an environment, stewardship initiatives can be perceived as secondary or even burdensome, leading to insufficient support from leadership. Ensuring that antimicrobial stewardship is recognized as a critical component of overall patient safety and quality of care requires strong advocacy and clear communication of its long-term benefits.

Patient expectations and misconceptions about antibiotics present another layer of complexity (Dikkatwar et al., 2024). Many patients expect or demand antibiotics for conditions that do not require them, such as viral infections. Educating patients about the appropriate use of antibiotics and the risks of overuse and misuse is essential, but changing public perception and behavior is a slow and ongoing process. Healthcare providers must balance patient satisfaction with the principles of stewardship, often navigating difficult conversations to avoid unnecessary antibiotic prescriptions.

In summary, antimicrobial stewardship within the global healthcare delivery system encounters numerous barriers that impede its effectiveness. These include disparities in resource availability, entrenched prescribing practices among healthcare providers, challenges in integrating data across systems, misalignments within organizational and system structures, and widespread misconceptions among patients about antibiotics. Tackling these issues demands a comprehensive strategy that encompasses continuous education for healthcare professionals, the establishment of advanced and reliable data systems, unwavering leadership commitment, and proactive public engagement to address misinformation. Successfully addressing these challenges is essential for enhancing stewardship initiatives and mitigating the escalating threat of antibiotic resistance, which poses a profound risk to global health security and sustainability.

9.5 Infection Prevention and Control: Anti-Microbial Stewardship in Action.

Infection prevention and control (IPC) is a fundamental aspect of healthcare service delivery that aims to protect patients, healthcare workers, and the community from the spread of infectious diseases (Tartari et al., 2021; Alhumaid et al., 2021; Billings et al., 2019).

Effective IPC practices are essential for reducing the incidence of healthcare-associated infections (HAIs), which can lead to serious complications, prolonged hospital stays, and increased healthcare costs. These practices include a combination of measures such as hand hygiene, the use of personal protective equipment (PPE), environmental cleaning, sterilization of medical equipment, and isolation protocols. By implementing robust IPC programs and the continued monitoring long-term, healthcare facilities can create safer environments, improve patient outcomes, and minimize the transmission of nosocomial infections.

Hand hygiene is the cornerstone of infection prevention and control (IPC) (Mouajou et al., 2022). Proper handwashing with soap and water or the use of alcohol-based hand sanitizers is one of the most effective ways to prevent the spread of pathogens from human-to-human. Healthcare workers must adhere to strict hand hygiene protocols before and after patient contact, after contact with potentially contaminated surfaces, and before performing any aseptic tasks. Ensuring compliance with hand hygiene guidelines requires continuous education, monitoring, and feedback. In addition, providing adequate hand hygiene amenities and supplies is critical to supporting this vital practice.

The use of personal protective equipment (PPE) is another key component of IPC. PPE includes items such as gloves, masks, gowns, and face shields, which provide a barrier against infectious agents. The appropriate selection and use of PPE depend on the nature of the patient interaction and the type of infection being managed. Training healthcare workers on the correct donning and doffing techniques is essential to prevent self-contamination and the spread of pathogens within healthcare settings. The COVID-19 pandemic has highlighted the importance of PPE in protecting healthcare workers and controlling the spread of respiratory infections, underscoring the need for adequate PPE supplies and proper usage protocols (Janson et. al., 2022; Liu, 2024)

Environmental cleaning and disinfection are critical for reducing the transmission of nosocomial infections. Pathogens (i.e., viruses and bacteria) are able survive on surfaces for extended periods, making regular cleaning and disinfection of patient rooms, common areas, and medical equipment essential. IPC programs must establish comprehensive cleaning protocols that specify the frequency and methods of cleaning, as well as the types of disinfectants to be used (Edward et. al., 2024). Training housekeeping staff and ensuring adherence to these protocols are crucial for maintaining a hygienic environment and preventing the spread of nosocomial infections.

Sterilization and disinfection of medical equipment are also vital to IPC (Thakur et al., 2024). Reusable medical instruments and devices must be properly cleaned, disinfected, or sterilized according to established guidelines to prevent cross-contamination between patients. Sterilization processes, such as autoclaving, and high-level disinfection methods are used to eliminate all forms of microbial life, including spores. Healthcare facilities must implement stringent protocols for the handling, processing, and storage of sterile equipment to ensure its safety and effectiveness.

Isolation protocols are necessary for managing patients with contagious infectious diseases (Blackwell, 2024). These protocols involve isolating infected patients to prevent the spread of pathogens to other patients, healthcare workers, and visitors. Different types of isolation, such as contact, droplet, and airborne precautions, are employed based on the mode of transmission of the infectious agent. Implementing isolation measures requires proper patient placement, use of PPE, and strict adherence to hand hygiene and environmental cleaning practices. Effective communication and coordination among healthcare team members are essential to successfully manage isolation cases.

In conclusion, infection prevention and control (IPC) are integral to maintaining the health, safety, and welfare of patients, healthcare workers, and the community. By adhering to rigorous hand-hygiene practices, utilizing appropriate PPE, ensuring thorough environmental cleaning, properly sterilizing medical equipment, and implementing isolation protocols for highly contagious infections, healthcare facilities can significantly reduce the risk of hospital-acquired infections and limit the spread of infectious diseases. Strong IPC programs, supported by continuous education, training, and compliance monitoring, are crucial for achieving these goals and fostering a culture of safety and accountability in the delivery of healthcare services.

9.6 The Future of Anti-Microbial Stewardship Around-the-Globe.

The future of antimicrobial stewardship around-the-globe is poised to be shaped by a combination of major technological advancements, policy initiatives, and evolving healthcare practices. With the growing threat of antibiotic resistance, which poses a significant public health risk, there is an increasing emphasis on optimizing the use of antimicrobials (Cocker et al., 2024). This involves not only prescribing the right antibiotics when necessary but also minimizing unnecessary usage. As such, the future will likely see a more robust implementation of stewardship programs across healthcare settings, including hospitals, outpatient clinics, and long-term care facilities. These programs will focus on improving diagnostic accuracy, enhancing infection control and prevention measures, and promoting the use of narrow-spectrum antibiotics to preserve the efficacy of broad-spectrum options (Williams et al., 2024).

Technological innovations, such as rapid diagnostic tests and advanced data analytics, will play a crucial role in the evolution of antimicrobial stewardship (Bonello et al., 2024). Rapid diagnostics can help healthcare providers quickly identify the pathogens causing infections and determine their susceptibility to various antibiotics, allowing for more targeted and effective treatment. Additionally, the integration of electronic health records (EHRs) and health information exchanges (HIEs)with decision-support tools can provide real-time alerts and recommendations to clinicians, ensuring adherence to empirically-derived evidence-based clinical practice guidelines (White et al., 2023). The use of big data analysis, machine learning (ML), and artificial intelligence (AI) can also enable the analysis of prescribing patterns and resistance trends, helping to refine stewardship strategies and track their impact over time (de la Lastra et al., 2024; Harry, 2024).

Public policy and regulatory frameworks will continue to be key drivers in advancing antimicrobial stewardship efforts (Cosgrove et al., 2023). The Centers for Disease Control and Prevention (CDC) and other governmental bodies at all levels of organization are likely to implement more stringent regulations and incentives to encourage the adoption of antimicrobial stewardship practices. This may include mandatory reporting of antibiotic use and resistance data, as well as financial incentives for healthcare facilities that demonstrate effective stewardship programs. Additionally, public education campaigns will be critical in raising awareness about the appropriate use of antibiotics among patients and the general public, thereby reducing the demand for unnecessary prescriptions (Parveen et al., 2022).

The future of antimicrobial stewardship in the global healthcare delivery system will also be shaped by a more collaborative approach across various sectors, including healthcare, agriculture, and veterinary medicine (Zinsstag et al., 2021). The *One Health* framework, which recognizes the inextricable interconnectedness of human, animal, and environmental health, will become increasingly important in addressing the issue of antibiotic resistance comprehensively. This integrated approach will involve coordinated efforts to monitor and control antibiotic use in agriculture, where overuse and misuse can contribute to the development of resistant bacteria that can spread to humans through food ingestion.

Overall, the future of global antimicrobial stewardship is likely to be characterized by a more proactive, data-driven, and multidisciplinary approach. By leveraging technological advancements, public policy measures, and collaborative efforts, the goal will be to preserve the efficacy of existing antibiotics, develop new treatment options, and ultimately safeguard the public's health against the growing threat of antibiotic resistance world-wide.

9.7 References

1. Ahmad, M., Khan, A. U. (2019) Global economic impact of antibiotic resistance: A review. *Journal of Global Antimicrobial Resistance* 19: 313-316.
2. Alhumaid, S., Al Mutair, A., Al Alawi, Z., et al. (2021) Knowledge of infection prevention and control among healthcare workers and factors influencing compliance: a systematic review. *Antimicrobial Resistance & Infection Control* 10(1): 86.
3. Aljeldah, M. M. (2022) Antimicrobial resistance and its spread is a global threat. *Antibiotics* 11(8): 1082.
4. Ancillotti, M., Eriksson, S., Veldwijk, J., et al. (2018) Public awareness and individual responsibility needed for judicious use of antibiotics: a qualitative study of public beliefs and perceptions. *BMC Public Health* 18: 1-9.
5. Årdal, C., Balasegaram, M., Laxminarayan, et al. (2020) Antibiotic development—economic, regulatory and societal challenges. *Nature Reviews Microbiology* 18(5): 267-274.
6. Bankar, N. J., Ugemuge, S., Ambad, et al. (2022). Implementation of antimicrobial stewardship in the healthcare setting. *Cureus* 14(7).
7. Barlam, T. F. (2021) The state of antibiotic stewardship programs in 2021: The perspective of an experienced steward. *Antimicrobial Stewardship & Healthcare Epidemiology* 1(1): e20.
8. Birgand, G., Dhar, P., Holmes, A. (2023) The threat of antimicrobial resistance in surgical care: the surgeon's role and ownership of antimicrobial stewardship. *British Journal of Surgery* 110(12): 1567-1569.
9. Billings, C., Bernard, H., Caffery, L., et al. (2019) Advancing the profession: An updated future-oriented competency model for professional development in infection prevention and control. *American Journal of Infection Control* 47(6): 602-614.
10. Blackwell, C. W. (2024) Application of the Centers for Disease Control and Prevention's isolation directives for patients exposed

to severe acute respiratory syndrome coronavirus 2. *The Journal for Nurse Practitioners* 20(4): 104962.

11. Bonello, M., Applegate, T. L., Badman, S., et al. (2024) The AMR Hub: a public–private partnership to overcome barriers to commercialisation and deliver antimicrobial stewardship innovations. *Microbiology Australia.*
12. Classen, D. C., Rhee, C., Dantes, R. B., et al. (2024) Healthcare-associated infections and conditions in the era of digital measurement. *Infection Control & Hospital Epidemiology* 45(1): 3-8.
13. Cocker, D., Birgand, G., Zhu, N., et al. (2024) Healthcare as a driver, reservoir and amplifier of antimicrobial resistance: opportunities for interventions. *Nature Reviews Microbiology* pp 1-14.
14. Conner, D. R. (1992) *Managing at the Speed of Change: How Resilient Managers Succeed and Prosper Where Others Fail.* Villard Books.
15. Cosgrove, S. E., Ahn, R., Dullabh, P., et al. (2024) Lessons Learned from a National Hospital Antibiotic Stewardship Implementation Project. *The Joint Commission Journal on Quality and Patient Safety* 50(6): 435-441.
16. Cosgrove, S. E., Srinivasan, A. (2023) Antibiotic stewardship: a decade of progress. *Infectious Disease Clinics* 37(4): 659-667.
17. de la Lastra, J. M. P., Wardell, S. J., Pal, T., et al. (2024) From Data to Decisions: Leveraging Artificial Intelligence and Machine Learning in Combating Antimicrobial Resistance–a Comprehensive Review. *Journal of Medical Systems* 48(1): 71.
18. Dhingra, S., Rahman, N. A. A., Peile, E., et al. (2020) Microbial resistance movements: an overview of global public health threats posed by antimicrobial resistance, and how best to counter. *Frontiers in Public Health* 8: 535668.
19. Dickman, S. L., Himmelstein, D. U., Woolhandler, S. (2017) Inequality and the health-care system in the USA. *The Lancet* 389(10077): 1431-1441.

20. Dikkatwar, M. S., Chand, S., Varghese, et al. (2024) Antimicrobial stewardship: smart approach to combat antibiotic resistance. *Anti-infective Agents* 22(4): 23-31.
21. Domer, G., Gallagher, T. M., Shahabzada, S., et al. (2021) Patient safety: preventing patient harm and building capacity for patient safety. *In* Contemporary Topics in Patient Safety-Volume 1. Intech Open.
22. Edward, M., John, W., Mahulu, E., et al. (2024) Challenges of Compliance with Infection Prevention and Control (IPC) Standard Procedures among Healthcare Workers: A Hospital-Based Cross-Sectional Study. *International Journal of Health Policy Planning* 3(1): 1-7.
23. Ferri, M., Ranucci, E., Romagnoli, P., et al. (2017) Antimicrobial resistance: A global emerging threat to public health systems. *Critical Reviews in Food Science and Nutrition* 57(13): 2857-2876.
24. Frieri, M., Kumar, K., Boutin, A. (2017) Antibiotic resistance. *Journal of Infection and Public Health* 10(4): 369-378.
25. Giamarellou, H., Galani, L., Karavasilis, T., et al. (2023) Antimicrobial stewardship in the hospital setting: a narrative review. *Antibiotics* 12(10): 1557.
26. Harry, A. (2024) Revolutionizing Healthcare: The Role of Artificial Intelligence in Antibiotic Stewardship and Resistance Management. *International Journal of Multidisciplinary Sciences and Arts* 3(1): 325-332.
27. Howard, A., Aston, S., Gerada, A., et al. (2024) Antimicrobial learning systems: an implementation blueprint for artificial intelligence to tackle antimicrobial resistance. *The Lancet Digital Health* 6(1): e79-e86.
28. Hwang, S., Kwon, K. T. (2021). Core elements for successful implementation of antimicrobial stewardship programs. *Infection & Chemotherapy* 53(3), 421.
29. Janson, D. J., Clift, B. C., Dhokia, V. (2022) PPE fit of healthcare workers during the COVID-19 pandemic. *Applied Ergonomics* 99: 103610.

30. Joshi, P., Bajpai, D. K. 2023 Antibiotic resistance: overuse of antibiotics, context and potential solutions. *IJTI* 1(2): viii-xv.
31. Kapadia, S. N., Abramson, E. L., Carter, E. J., et al. (2018) The expanding role of antimicrobial stewardship programs in hospitals in the United States: lessons learned from a multisite qualitative study. *The Joint Commission Journal on Quality and Patient Safety* 44(2): 68-74.
32. Lane, M. A., Hays, A. J., Newland, H., et al. (2019) Development of an antimicrobial stewardship program in an integrated health-care system. *American Journal of Health-System Pharmacy* 76(1): 34-43.
33. Laxminarayan, R., Matsoso, P., Pant, S., et al. (2016) Access to effective antimicrobials: a worldwide challenge. *The Lancet* 387(10014): 168-175.
34. Lim, J. M., Singh, S. R., Duong, M. C., et al. (2020) Impact of national interventions to promote responsible antibiotic use: a systematic review. *Journal of Antimicrobial Chemotherapy* 75(1): 14-29.
35. Liu, C. (2024) PPE-Based Strategies For COVID-19 Prevention Amongst Healthcare Workers. *Highlights in Science, Engineering and Technology* 102: 709-717.
36. Logan, A. Y., Williamson, J. E., Reinke, et al. (2019) Establishing an antimicrobial stewardship collaborative across a large, diverse health care system. *The Joint Commission Journal on Quality and Patient Safety* 45(9): 591-599.
37. Mouajou, V., Adams, K., DeLisle, G., et al. (2022). Hand hygiene compliance in the prevention of hospital-acquired infections: a systematic review. *Journal of Hospital Infection* 119: 33-48.
38. Ohlsen, J. T., Søfteland, E., Akselsen, P. E., et al. (2024). Rapid response systems, antibiotic stewardship and medication reconciliation: a scoping review on implementation factors, activities and outcomes. *BMJ Quality & Safety*.

39. Owens Jr, R. C. (2008) Antimicrobial stewardship: concepts and strategies in the 21st century. *Diagnostic Microbiology and Infectious Disease* 61(1): 110-128.
40. Parveen, S., Garzon-Orjuela, N., Amin, D., et al. (2022) Public health interventions to improve antimicrobial resistance awareness and behavioural change associated with antimicrobial use: a systematic review exploring the use of social media. *Antibiotics* 11(5): 669.
41. Patel, K., Rushefsky, M. E. (2019) *Healthcare politics and policy in America.* Routledge.
42. Powers, J. H. (2024). Antimicrobial stewardship. *BMJ* 385.
43. Sipahi, O. R. (2008) Economics of antibiotic resistance. *Expert Review of Anti-infective Therapy* 6(4): 523-539.
44. Tamma, P. D., Cosgrove, S. E. (2011) Antimicrobial stewardship. *Infectious Disease Clinics* 25(1): 245-260.
45. Tartari, E., Tomczyk, S., Pires, D., et al. (2021). Implementation of the infection prevention and control core components at the national level: a global situational analysis. *Journal of Hospital Infection* 108: 94-103.
46. Teixeira, C. F. D. S., Soares, C. M., Souza, E. A., et al. (2020) The health of healthcare professionals coping with the Covid-19 pandemic. Ciencia & Saude Coletiva 25: 3465-3474.
47. Thakur, H., Rao, R. (2024) Emphasis of infection prevention and control: a review. *J Popul Therap Clin Pharmacol* 31: 2238-49.
48. Van Dort, B. A. (2024) *An in-depth exploration of antimicrobial stewardship: The interplay between people and systems.* University of Sydney Doctoral Dissertation.
49. Vaughn, V. M., Krein, S. L., Hersh, A. L., et al. (2024) Excellence in Antibiotic Stewardship: A Mixed-Methods Study Comparing High-, Medium-, and Low-Performing Hospitals. *Clinical Infectious Diseases* 78(6): 1412-1424.
50. Warner, M. A., Apfelbaum, J. L. (2015) The perioperative surgical home: a response to a presumed burning platform or a

thoughtful expansion of anesthesiology? *Anesthesia & Analgesia* 120(5): 1149-1151.

51. Weinstein, R. Buckel, W. R., Stenehjem, E. A., et al. (2022) Harnessing the power of health systems and networks for antimicrobial stewardship. *Clinical Infectious Diseases* 75(11): 2038-2044.
52. White, N. M., Carter, H. E., Kularatna, S., et al. (2023) Evaluating the costs and consequences of computerized clinical decision support systems in hospitals: a scoping review and recommendations for future practice. *Journal of the American Medical Informatics Association* 30(6): 1205-1218.
53. Williams, S. B., McCaffrey, P., Reynoso, D., et al. (2024) Implementation of a High-Value, Evidence-Based Care Program: Impact and Opportunities for Learning Organizations. *Journal of Healthcare Management* 69(4): 296-308.
54. Zinsstag, J., Schelling, E., Crump, L., et al. (eds.) (2021) *One Health: the theory and practice of integrated health approaches.* CABI.

10.0

The Future of Sustainable Healthcare Stewardship.

10.1 The Future of Healthcare Stewardship in 21st Century Global Healthcare Delivery Systems.

The future of healthcare stewardship in the 21st century global healthcare delivery system will be influenced by various factors, such as major technological advancements, demographic shifts, and changing societal expectations. As healthcare delivery systems worldwide face the mounting challenges of aging populations, rising healthcare costs, and the constant threat of pandemics, responsible, effective, empirically supported, and evidence-based clinical practice guidelines and healthcare management will become ever more essential for ensuring the health, wellness, well-being, and resilience of future generations.

The integration of innovative technologies, such as artificial intelligence (AI), machine learning (ML), telemedicine, and big data analytics, offers significant potential to enhance the efficiency and effectiveness of global healthcare services delivery (Maleki-Vamosfaderani et al., 2024). By leveraging these tools, global healthcare delivery systems can improve equitable resource allocation, streamline operations, and personalize patient care, ultimately leading to better health outcomes and more environmentally sustainable healthcare delivery systems.

In the coming years, healthcare stewardship will need to increasingly focus on population health management, people-centered care, and value-based approaches to resource utilization, with the goal of improving quality and reducing the overall cost of care. This shift requires a more proactive approach, focusing on health prevention, promotion, protection, and preparedness with early detection and management of both communicable and non-communicable chronic diseases. Healthy public policies and practices, like the American Heart Association's *Life's Essential 8* program (Lloyd-Jones et al., 2022; Ma et al., 2023) that promotes pursuing and maintaining a healthy cardiovascular system with an overarching goal of increasing an individual's health span meeting up with the individual's lifespan, will be an essential component of this global strategy. Additionally, addressing upstream drivers of healthcare utilization including the health-related social needs (HRSNs) and other economic, environmental and commercial determinants of health will be critical for reducing health disparities, improving health literacy, and ensuring equitable access to primary health care. Effective stewardship in this context means creating a healthcare delivery system that actively produces positive health, optimizes wellness, enhances well-being, and develops resilience across all population groups.

Another significant aspect of the future of healthcare stewardship is the increasing need for international collaboration, coordination, and cooperation. Existential global health challenges, such as emerging infectious disease outbreaks and the effects of the Anthropocene climate change on health status, will require intensive efforts from countries and healthcare delivery systems worldwide to manage and even reverse these threats to human existence. Strengthening global health governance, enhancing data sharing, analytics, and interoperability, and fostering cross-border partnerships will be essential for addressing these issues efficiently, equitably, and effectively. Furthermore, the design and implementation of robust healthy public policies that are adaptable to diverse contexts will help ensure that global healthcare delivery systems can respond swiftly and efficiently. By fostering cultural humility and shared responsibility, the global healthcare community can work together to build resilient healthcare delivery systems capable of withstanding these and any other future challenges (e.g., economic inequities and institutional bias and racism).

Finally, the moral and ethical dimensions of healthcare stewardship will continue to play a pivotal role in shaping the future of healthcare services delivery world-wide. As new global technologies and treatments emerge at exponential speed, moral and ethical considerations surrounding resource allocation, patient autonomy, and data access and privacy will become increasingly complex. Ensuring that healthcare decisions are made transparently and inclusively, with input from diverse healthcare stakeholders, will be critical for preserving public trust and achieving equitable health outcomes. Additionally, fostering a culture of continuous learning and improvement within healthcare delivery systems and individual healthcare organizations around-the-world, will be essential for adapting to changing circumstances and advancing the goals of healthcare stewardship. By prioritizing moral and ethical principles and embracing creativity and innovation, global healthcare delivery systems can navigate the challenges that are sure to come in the not-so-distant future.

10.2 Leveraging Healthcare Stewardship to Attain Excellence in Practice and Optimal Outcomes.

To achieve best evidence-based clinical practices and subsequently successful health outcomes through healthcare stewardship, it is essential to implement a variety of empirically-derived and evidence-based strategies. These practices ensure the efficient, equitable, and effective use of healthcare resources world-wide. Below are examples of the key approaches discussed throughout this book to achieve healthcare stewardship.

1. *Evidence-Based Clinical Practices.*
    a) *Clinical Practice Guidelines*: Designing and implementing empirically-derived, evidence-based clinical practice guidelines is crucial for standardizing all clinical care and reducing variability across healthcare delivery settings. These guidelines ensure that all patients receive high-quality, cost-effective interventions by providing clear, research-backed recommendations for management and treatment of disease. By enhancing reliability and consistency in healthcare services, these guidelines help improve patient outcomes and optimize resource use, ultimately leading to a more effective and efficient global healthcare delivery system.
    b) *Continuous Learning*: Encouraging continuous and comprehensive professional development and training for all clinical providers and staff is essential to keep pace with the latest evidence-based clinical practices. By adopting the principles and practices of a learning health system, healthcare delivery systems can ensure that their workforce remains current with advancements in medical knowledge and technology, decreasing the time it takes to go from the bench to the bedside to the population to society. This ongoing education helps healthcare providers deliver the highest quality of care, fosters a culture of continuous improvement, and enhances overall patient outcomes by integrating the most recent and effective practices into everyday care focusing on healthcare stewardship principles.
    c) *Research and Innovation*: Investing in, designing, and implementing empirically-derived healthcare research and development is crucial for advancing positive health outcomes. By leveraging advanced and innovative science and technol-

ogies, healthcare delivery systems world-wide can develop new treatments, diagnostic tools, and proactive measures that address current challenges and improve patient care. This commitment to research ensures that medical practices are based on the latest evidence and available technologies, leading to more cost-effective interventions, enhanced overall health outcomes, and healthcare stewardship.

2. *Holistic, Integrated, Comprehensive and Coordinated Clinical Care.*
   a) Holistic, Integrated, and Comprehensive Clinical Care: An interdisciplinary and multisectoral healthcare team, encompassing primary health care with advanced primary care services, specialty medical services, essential public health services, and community-based services, collaborates to provide comprehensive care for individuals and the broader community. By employing a health equity and culturally competent approach, the team addresses the physical, psychological, and social aspects of health, striving to achieve positive health outcomes, optimize wellness, enhance well-being, and develop resilience. This integrated approach ensures that healthcare services delivery is tailored to the diverse needs of patients, fostering a more inclusive and effective healthcare delivery system while simultaneously monitoring for appropriate use of healthcare resources.
   b) *Care Coordination*: Developing and implementing care coordination frameworks is essential to ensure that all patients, especially those with chronic conditions, have access to and receive the necessary navigation assistance during transitions of care across different healthcare settings. These frameworks facilitate smooth handoffs between providers, enhance communication, and streamline processes, ensuring continuity of care. By providing structured support and guidance, care coordination frameworks help patients man-

age their chronic conditions more effectively, reduce gaps in care, and improve overall health outcomes.

c) *Multidisciplinary Teams*: Employing multidisciplinary teams to deliver evidence-based, value-added healthcare services ensures that clinical care is both comprehensive and responsive to diverse patient needs. By integrating cultural humility and empathy into their approach, these teams address health disparities and work to close gaps in clinical care. This collaborative model emphasizes healthcare stewardship through enhancing the effectiveness of interventions and promoting equitable, people-centered care across various healthcare settings.

d) *People-Centered Medical Homes*: The people-centered medical home (PCMH) model of care is founded on principles and practices that include a whole-person orientation and team-based care. It emphasizes respect for patient preferences, needs, and values, delivered with cultural humility and empathy. The PCMH integrates healthcare services and functions, ensures seamless transitions of care, enhances access and navigation to care, and fosters open, bidirectional communication channels. It provides evidence-based services, engages in continuous quality improvement, empowers patients through self-management, addresses all determinants of health, and operates within a learning health system framework.

3. *Health Information Technology.*

a) *Electronic Health Records (EHRs)*: Leveraging electronic health records (EHRs) enhances the accuracy, accessibility, and efficiency of both clinical and non-clinical patient information. This technology streamlines the management of secure data, facilitating better coordination and continuity of care during transitions of care within the healthcare delivery system. By providing a comprehensive and up-to-date view of patient information, EHRs improve communication among providers and support seamless healthcare services delivery across different healthcare settings reducing wasteful and duplicative clinical care.

b) *Telemedicine*: Expanding telemedicine can enhance access to care, especially in underserved and remote areas, and provide timely, convenient consultations for both physical and behavioral health issues. This growth requires improved broadband internet coverage and access to smart technology for the most vulnerable and marginalized patients.

c) *Data Analytics*: Utilizing big data and data analytics allows healthcare delivery systems to identify trends, predict outcomes, and guide decision-making processes with greater precision. By analyzing vast amounts of data from diverse sources, such as patient records, clinical outcomes, and population health metrics, these tools uncover patterns and insights that inform targeted interventions and strategic planning. This approach enhances the ability to anticipate future health trends, optimize resource allocation, and improve patient care, ultimately leading to more effective and personalized healthcare solutions.

d) *Interoperability*: Improving the interoperability of health information systems and technology, along with expanding the use of Health Information Exchanges (HIEs), has become essential for a healthcare stewardship-influenced modern healthcare system. These advancements enable the seamless sharing of patient information across various healthcare providers and systems. Interoperability ensures that various electronic health records (EHR) systems can communicate and exchange data accurately, efficiently, and securely, regardless of the software used. HIEs act as centralized platforms or networks that enable the transfer of health information among diverse healthcare organizations, including hospitals, clinics, pharmacies, and public health agencies. This connectivity enhances the coordination of care, reduces the likelihood of medical errors, and improves clinical outcomes by providing healthcare professionals with timely and comprehensive access to patient records. By fostering a more integrated healthcare ecosystem, interoperability and HIEs also empower patients with better access to their health information, thereby promoting patient engagement and enabling more informed healthcare decisions.

e) *Population Health Management (PHM) Platforms*: Implementing a Population Health Management (PHM) platform serves as a comprehensive tool designed to improve health outcomes and optimize care for entire populations. These platforms integrate data from various sources, including electronic health records (EHRs), claims data, and patient surveys, to provide a holistic view of population health. By leveraging advanced analytics and data visualization, PHM platforms identify trends, assess risk factors, and stratify patient populations based on their health needs. This enables healthcare delivery systems to implement targeted interventions, manage chronic conditions more effectively, and promote proactive care. Additionally, PHM platforms facilitate care coordination among providers, streamline workflows, and support value-based care initiatives by monitoring performance metrics and patient outcomes. Ultimately, PHM platforms enhance the ability of healthcare delivery systems to deliver personalized, efficient, and proactive healthcare services, improving overall population health and reducing healthcare costs.

4. *Quality Improvement Initiatives.*

a) *Performance Measurement*: Implementing performance measurement systems is essential for monitoring and evaluating the quality of care provided. By utilizing key performance indicators such as patient outcomes, readmission rates, emergency department (ED) volume, and patient satisfaction, healthcare organizations can assess the effectiveness of their services and identify areas for improvement. These measurements help ensure that clinical healthcare standards are met, facilitate continuous quality improvement, and ultimately lead to better patient outcomes and enhanced overall healthcare services delivery.

b) *Continuous Improvement*: Fostering a culture of continuous improvement involves consistently evaluating and refining policies, processes, and practices to elevate the quality, safety, and efficiency of care provided. By routinely reviewing workflows, seeking feedback from staff and patients, and implementing data-driven changes, healthcare delivery systems can adapt to evolving needs, challenges, and efficiencies at scale. This proactive approach not only enhances operational effectiveness but also ensures that healthcare services delivery remains people-centered and responsive. Embracing a mindset of ongoing refinement helps cultivate an environment where excellence is pursued relentlessly, benefiting both healthcare providers and those they serve.

c) *Accreditation and Certification*: Pursuing accreditation and certification from internationally- and nationally-recognized healthcare organizations is essential for ensuring that healthcare infrastructure adheres to the highest standards of care. These prestigious endorsements validate that a facility meets rigorous criteria for quality, safety, and operational excellence. By engaging in this process, healthcare providers demonstrate their commitment to maintaining superior care practices, fostering trust with patients and stakeholders. Achieving and upholding these credentials not only enhances a facility's reputation but also drives continuous improvement, ultimately leading to better patient outcomes and more effective healthcare delivery.

5. *Patient Engagement and Education.*
   a) *Shared Decision-Making*: Involving patients in their own care decisions is crucial for ensuring they receive personalized and effective diagnostics and treatments. By providing patients with comprehensive information and supportive evidence-based resources, healthcare providers empower them to make informed choices about their health and healthcare. This collaborative approach fosters greater engagement, enhances patient satisfaction and the patient experience, and leads to better adherence to treatment plans. When patients are actively involved in their care, they are more likely to feel respected and valued, which can significantly improve their overall health outcomes and experience with the healthcare system.
   b) *Health Literacy*: Improving health literacy is essential for empowering patients to take control of their health by educating them about their conditions, treatment options, and proactive measures. By providing clear, accessible appropriate health information and using effective communication strategies, healthcare providers help patients understand their health preferences, needs, and values when making shared- informed decisions. This enhanced understanding not only supports better management of health conditions but also promotes engagement in proactive care. As patients become more knowledgeable about their health, they are better equipped to collaborate with their providers, leading to improved outcomes and a more effective healthcare experience.
   c) *Patient Feedback*: Collecting and acting on patient feedback is vital for identifying areas for improvement and ensuring that healthcare services delivery aligns with patients' cultural needs, preferences, and values. By actively seeking and responding to patient input, healthcare providers can gain valuable insights into the effectiveness of their healthcare services and make necessary adjustments to better meet

diverse patient needs. This approach fosters a more personalized and respectful healthcare environment, enhances the patient experience, and builds trust and truthfulness between patients and providers. Addressing feedback in a thoughtful and responsive manner not only improves care quality but also helps create a more inclusive and responsive healthcare experience.

6. *Resource Management.*
   a) *Cost-Effectiveness*: Prioritizing cost-effective interventions ensures that healthcare resources are allocated in a way that maximizes health outcomes relative to their cost. By leveraging health technology assessments, healthcare providers can evaluate the effectiveness and economic impact of various interventions, guiding decisions toward those that offer the greatest benefit at the most reasonable cost. This approach not only helps in optimizing resource use but also supports the delivery of high-quality care within budgetary constraints. Implementing cost-effective strategies allows healthcare systems to enhance patient outcomes while maintaining financial sustainability.
   b) *Waste Reduction*: Identifying and eliminating wasteful practices, such as unnecessary tests and procedures, is crucial for enhancing efficiency and reducing costs in healthcare. By scrutinizing and streamlining processes, healthcare providers can avoid redundant or non-beneficial activities that do not contribute to patient care, thereby optimizing resource use. This focus on minimizing waste not only helps in cutting costs but also improves overall care delivery by ensuring that resources are directed toward activities that offer real value. Streamlining practices leads to more efficient operations, better patient outcomes, and a more sustainable healthcare system.

c) *Sustainable Practices*: Implementing environmentally-sustainable practices is essential for reducing the healthcare delivery system's environmental footprint and promoting long-term sustainability. By adopting eco-friendly measures, such as reducing all forms of healthcare related-waste, conserving energy, and using sustainable materials, healthcare facilities can significantly lessen their environmental impact. These practices not only contribute to a healthier planet but also align with broader goals of sustainability, ultimately benefiting both current and future generations. Adopting green initiatives in healthcare promotes a holistic, integrated, and comprehensive stewardship approach to health, wellness, well-being, and resiliency recognizing that individual, family, and community health is closely linked to environmental health.

d) *Medical Supply Chain Management*: Regular assessments of medical supply chain activity are crucial for identifying and addressing early bottlenecks in structures and processes. By systematically evaluating supply chain performance, healthcare delivery systems can pinpoint inefficiencies, delays, or obstacles before they escalate into more significant issues. This proactive approach enables timely interventions to streamline operations, improve workflow, and ensure the consistent availability of necessary resources. Effective medical supply chain management not only enhances overall efficiency but also supports uninterrupted patient care and optimal operational performance.

7. *Health Equity and Access.*

a) *Equitable Resource Allocation*: Ensuring equitable distribution of healthcare resources is vital for reducing health disparities, particularly among underserved and vulnerable populations. By prioritizing these groups, healthcare delivery systems can address gaps in access and care, providing essential healthcare services where they are most needed. This approach promotes fairness and helps to rectify systemic imbalances, ensuring that all individuals receive appropriate and timely care. Focusing on equitable resource allocation contributes to a more inclusive healthcare delivery system, ultimately improving overall public health and reducing disparities.

b) *Access and Navigation to Care*: Improving access to empirically-derived evidence-based clinical care requires implementing healthy public policies and programs that address upstream barriers such as geographic location, socio-economic status, and cultural differences. By developing initiatives that specifically target these challenges, healthcare delivery systems can enhance navigation and ensure that quality care reaches all individuals, regardless of their circumstances. Healthy public policies that expand healthcare coverage, support telemedicine, and provide culturally competent care are essential for overcoming these barriers. This comprehensive approach promotes equitable access to healthcare services, ultimately leading to better health outcomes and a more inclusive healthcare environment.

c) *Community Health Initiatives*: Developing and supporting community health initiatives that address all determinants of health requires close collaboration with community-based organizations and community-oriented primary health care. By working together, these entities can create comprehensive strategies that tackle the social, economic, environmental, and commercial factors influencing health status. Engaging with local communities ensures that initiatives are tailored to specific needs and conditions, fostering more effective and sustainable health solutions. This collaborative approach not only enhances the reach and impact of health programs but also empowers communities to actively participate in improving their own health, wellness, well-being, and resilience.

d) *Community-based Organizations*: Implementing closed-loop referrals for upstream, non-clinical determinates of health and fully engaging with cultural humility and empathy are crucial for providing comprehensive and effective care. For instance, programs like *Food as Medicine* and *Non-Emergent Medical Transportation (NEMT)* address specific patient needs beyond traditional clinical settings, ensuring that individuals receive comprehensive support. By approaching these healthcare services with cultural humility, healthcare providers acknowledge and respect diverse backgrounds, fostering trust and truthfulness and improving patient engagement. This empathetic and integrated approach helps bridge gaps in healthcare services, enhances patient outcomes, and supports overall health, wellness, well-being, and resilience by addressing both medical and social determinants of health.

8. *Governance, Leadership, and Accountability.*
   a) *Transparent Decision-Making*: Ensuring that healthcare decisions are made transparently and involve input from a diverse array of stakeholders is essential for fostering trust and truthfulness and achieving balanced outcomes. By incorporating perspectives from patients, providers, and policymakers, healthcare delivery systems can address the needs and concerns of all parties involved, leading to more informed and equitable decisions. This inclusive approach not only enhances accountability and transparency but also promotes collaboration and shared responsibility in shaping effective healthcare policies and practices. Engaging all relevant stakeholders ensures that decisions reflect a comprehensive understanding of the challenges and opportunities within the healthcare system.

b) *Accountability Mechanisms*: Implementing accountability mechanisms is crucial for ensuring that healthcare providers and healthcare delivery systems adhere to established empirically-derived evidence-based best clinical practices and standards of healthcare stewardship. By setting clear performance benchmarks and regularly monitoring adherence, these mechanisms hold healthcare providers accountable for their actions and outcomes. This includes conducting audits, implementing performance reviews, and establishing transparent reporting systems (e.g., dashboards and scorecards). Such measures not only promote consistent delivery of high-quality care but also foster a culture of continuous improvement, where healthcare providers are motivated to maintain and exceed established standards. Ensuring accountability helps enhance patient safety, optimize care quality, and build trust within and outside of the healthcare delivery system.
c) *Policy and Regulation*: Developing and enforcing policies and regulations that promote high standards of care, protect patient rights, and ensure the responsible use of healthcare resources is essential for maintaining a healthcare stewardship-centric best-in-class healthcare delivery system. These policies provide clear evidence-based guidelines and accountability, ensuring that care is delivered consistently and ethically while safeguarding patient health, safety, and welfare. Regulations that address healthcare resource allocation and management help to prevent waste and misuse, promoting efficient and effective care. By establishing and upholding these standards, healthcare delivery systems can enhance patient safety, uphold ethical practices, and optimize the use of resources, ultimately leading to improved health outcomes and overall healthcare delivery system integrity.
d) *Trust and Truthfulness*: Leadership that prioritizes factual and verifiable information fosters full engagement among

all healthcare stakeholders by promoting transparency, trust, and truthfulness. When leaders base their decisions and communications on accurate data, they facilitate multi-directional communication, enabling open dialogue between healthcare providers, patients, policymakers, and other key parties. This evidence-based approach ensures that all stakeholders are informed and aligned, leading to more effective collaboration and informed decision-making. By upholding a commitment to factual accuracy, leaders can drive informed, constructive interactions and support a cohesive, responsive healthcare environment focused on the principles and practices of healthcare stewardship.

In summary, achieving empirically-derived evidence-based best clinical practice and successful positive outcomes in healthcare stewardship requires a multi-layered approach that integrates in a healthcare delivery system both horizontally and vertically focusing on evidence-based practices, coordinated care, advanced health information technology, continuous quality improvement, population health management, patient engagement, efficient resource management, health equity, and robust governance. By focusing on these foundational areas, global healthcare delivery systems can enhance the quality, efficiency, and equity of care, ultimately leading to better health outcomes for both individuals and the community-at large.

10.3 References

1. Lloyd-Jones, D. M., Ning, H., Labarthe, D., et al. (2022) Status of cardiovascular health in US adults and children using the American Heart Association's new "Life's Essential 8" metrics: prevalence estimates from the National Health and Nutrition Examination Survey (NHANES), 2013 through 2018. *Circulation* 146(11): 822-835.

2. Ma, H., Wang, X., Xue, Q., et al. (2023) Cardiovascular health and life expectancy among adults in the United States. *Circulation* 147(15): 1137-1146.
3. Maleki-Varnosfaderani, S., Forouzanfar, M. (2024) The role of AI in hospitals and clinics: transforming healthcare in the 21st century. *Bioengineering* 11(4): 337.